FORTRESS • 72

GERMAN V-WEAPON SITES 1943–45

STEVEN J ZALOGA   ILLUSTRATED BY HUGH JOHNSON & CHRIS TAYLOR

Series editors Marcus Cowper and Nikolai Bogdanovic

First published in Great Britain in 2007 by Osprey Publishing,
Midland House, West Way, Botley, Oxford OX2 0PH, United Kingdom
443 Park Avenue South, New York, NY 10016, USA
Email: info@ospreypublishing.com

A CIP catalogue record for this book is available from the British Library.

ISBN 978 184603 247 9

Editorial by Ilios Publishing, Oxford, UK (www.iliospublishing.com)
Page layout by Ken Vail Graphic Design, Cambridge, UK (kvgd.com)
Typeset in Sabon and Myriad Pro
Index by Glyn Sutcliffe
Maps by Map Studio Ltd, Romsey, UK
Originated by PDQ Digital Media Solutions Ltd, Bungay, UK
Printed and bound in China through Bookbuilders

07 08 09 10 11   10 9 8 7 6 5 4 3 2 1

FOR A CATALOGUE OF ALL BOOKS PUBLISHED BY OSPREY MILITARY
AND AVIATION PLEASE CONTACT:

NORTH AMERICA
Osprey Direct, c/o Random House Distribution Center, 400 Hahn Road,
Westminster, MD 21157
Email: info@ospreydirect.com

ALL OTHER REGIONS
Osprey Direct UK, PO Box 140, Wellingborough,
Northants, NN8 2FA, UK
Email: info@ospreydirect.co.uk

www.ospreypublishing.com

THE FORTRESS STUDY GROUP (FSG)

The object of the FSG is to advance the education of the public in the
study of all aspects of fortifications and their armaments, especially
works constructed to mount or resist artillery. The FSG holds an annual
conference in September over a long weekend with visits and evening
lectures, an annual tour abroad lasting about eight days, and an annual
Members' Day.

The FSG journal FORT is published annually, and its newsletter Casemate
is published three times a year. Membership is international. For further
details, please contact:

The Secretary, c/o 6 Lanark Place, London W9 1BS, UK
website: www.fsgfort.com

CONTENTS

GERMAN V-WEAPON SITES 1943–45

INTRODUCTION

The failure of the Luftwaffe in its attacks against Britain and the rising crescendo of RAF bomber attacks against Germany convinced Hitler in 1943 to substitute exotic new "Vengeance" weapons (*Vertgeltungswaffen*) to retaliate against London. The short range of these early missiles inevitably meant that they would be launched from near the British coast and so within the striking range of Allied bombers. There was considerable controversy how to base the missiles to make them the most survivable. Many Wehrmacht commanders favored mobile missile bases, but Hitler preferred heavy bunkers patterned after the impregnable U-boat bunkers. Besides the V-1 and V-2 missiles, other novel weapons were added to the arsenal, including the V-3 Tausenfüßler (millipede), a unique multi-stage artillery weapon capable of reaching London from the French coast. Critical British intelligence successes led to the discovery of the missile program months before they were ready for combat, and a pre-emptive air campaign was launched against the Crossbow

The Wehrmacht remained torn between mobile and fixed basing for its new secret weapons. The artillery branch, which controlled the V-2 ballistic missile, favored mobile basing using simple pad launchers like those seen here at Test Stand X at Peenemünde during training exercises for the experimental Batterie 444 in 1944. (MHI)

sites located in France in the autumn and winter of 1943. This derailed the original German scheme to start the attacks in December 1943 and forced the adoption of new basing modes for the V-weapons. Although the new sites proved to be less vulnerable to air attack than the initial heavy Crossbow sites (Crossbow was the codename for the British intelligence committee responsible for uncovering German V-weapon programs), they were also considerably less efficient and the V-weapons failed to have any major impact on the course of the war. In spite of their feeble results, the V-weapons were the ancestors of the Cold War's awesome nuclear missiles and their launch sites served as a guide for later missile launch complexes. The lessons of the first missile campaign were not forgotten, and the V-2 served as inspiration to the infamous Scud missile so prominent in wars of the Middle East in the last two decades of the 20th century.

THE V-WEAPON PROGRAMS

The German Army sponsored a ballistic missile program in the late 1930s as a form of long-range artillery. The intention was to develop a weapon capable of delivering a one-ton payload to a range ten-times that of the World War I Paris gun, roughly 165 miles (270km). The A-4 missile program was officially initiated in 1936, but the technology was so radical that a series of sub-scale missiles had to be designed and tested before the full-scale missile could be developed. These experimental missiles were launched from a secret test facility at Peenemünde on an isolated peninsula in the Baltic starting in 1938. The first full-size A-4 missile was completed in February 1942 but the first attempted test launch in March 1942 failed. The fourth test flight, on October 3, 1942, finally succeeded, but the design was far from mature and test launches continued through 1943 to make the A-4 suitable for combat use. This missile is better known by its later propaganda designation as the V-2 (Vertgeltungswaffe-2: Retaliation weapon-2).

In the summer of 1942, the German Army sought Hitler's approval to begin preparing for the mass-production of the A-4 missile, a major issue due to the enormous cost of the program, which was also likely to impact German aircraft production. The German generals promised that the new missile would succeed where the Luftwaffe had failed in the 1940 Battle of Britain. A storm of missiles would rain down on London, knocking Britain out of the

The FZG-76 cruise missile used a Walter steam catapult to get up enough speed for its Argus pulse-jet engine to ignite. This is an early test version of the launch system at the main experimental range at Peenemünde in the autumn of 1943. (NARA)

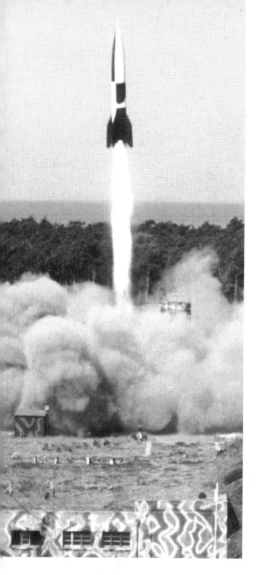

An early test example of the A-4 ballistic missile lifts off from Test Stand VII at Peenemünde in 1943. (NARA)

war. Hitler had been ambivalent about the missile program, but the growing ferocity of British bomber missions over Germany changed his mind.

The Luftwaffe's strategic bomber program had continued to stumble and there was little certainty that it would succeed. Although the Luftwaffe had rebuffed earlier attempts by aircraft companies to develop long-range strike missiles, the Army's campaign to take away precious production resources to build their missile was enough to lead to a rival Luftwaffe missile program. Instead of a ballistic missile, the Luftwaffe selected a cruise missile design offered by Fieseler, the Fi-103. The Luftwaffe gave it the cover-name FZG-76 (Flakzielgerat 76), linking it to the innocuous Argus FZG-43 target drone. It would be better known in later years as the V-1. In contrast to the expensive A-4 ballistic missile, the FZG-76 was designed to be cheap to build and simple to operate, using a rudimentary pulse-jet engine and a small and simple airframe that could be manufactured by any aircraft plant. The design was so simple that test examples began to fly by October 1942. Hitler recognized that the FZG-76 program was an inexpensive alternative to the much riskier A-4 missile program, and both programs were allowed to continue in parallel through 1943.

Following the fall of France in 1940, the German Army deployed long-range artillery on the Pas-de-Calais to support the intended invasion of England. Although they could reach as far as some coastal cities such as Dover, they could not reach much beyond due to the limits of conventional artillery technology. In 1942, Eisenbahn-Artillerie Batterie 725 near Calais was assigned the task of testing a new extended-range rocket-assisted artillery projectile from one of its 280mm K5(e) guns that were intended to reach London. The test was a failure when the enormous propellant charge ripped the barrel apart. Although efforts continued to develop long-range projectiles, a more promising technology was also being explored by the Röchling plant, called the Hochdruckpumpe (HDP: high pressure pump), or Tausenfüßler (Millipede). Instead of using a single propellant charge at the breech of the gun, the pump gun used a sequence of smaller charges located in small chambers along the barrel's 127m length. These were electrically fired as the projectile passed down the barrel, imparting energy more efficiently than a single charge. The aim was to develop a weapon capable of firing a 140kg projectile to a range of 165km. While this projectile was not as large as the warhead in a V-1 or V-2 missile, the presumption was that the low cost and volume of fire would make up for the relatively small payload. A sub-scale 20mm prototype was tested at a proving ground in Misdroy (now Miedzydroje, Poland) in April–May 1943, and the project attracted Hitler's attention. In August 1943, he authorized the construction of a 50-gun HDP battery in France to supplement the missile campaign against London. This gun battery would have a theoretical rate of fire of one shot per tube every minute, or 600 rounds per hour, and 20,000 rounds per month. Although the rounds were much smaller than the missiles, the sheer volume of fire was enough to excite Hitler's enthusiastic support. The full-scale prototype of the weapon was completed at the Wehrmacht's Hillersleben artillery proving ground in October 1943.

DESIGN AND DEVELOPMENT

During the design of the A-4 ballistic missile in 1941–42, the engineers began discussions about the possible launch configurations for the weapon. From a strictly technical standpoint, a fixed site was preferred for many reasons. To begin with, the A-4 missile was extremely complicated, requiring a substantial amount of test equipment to monitor the missile subsystems prior to launch. In addition, the A-4 used liquid oxygen (LOX) as the oxidizer for its fuel, and this chemical had to be maintained at super-cold temperatures using elaborate refrigeration and insulation techniques that were easier to undertake at a fixed site than at a mobile field site. Indeed, the liquid oxygen oxidizer, codenamed A-stoff by the Germans, would prove to be the main bottleneck in the combat launches of the A-4 missile. German industrial facilities produced only about 155 tons of LOX daily plus a further 60 tons in the occupied countries. Although the fueling operation for an A-4 missile consumed about 4.7 tons, on average it took about 15 tons per missile launch as about 5 tons of LOX boiled off during transit from the factory to the field. This implied that all of Europe produced only enough LOX to launch an average of 14 A-4 missiles daily with the assumption that all the LOX was available to the missile program, which of course was not the case due to industry and military requirements. One solution was to substantially increase the production capacity for LOX, but there would still be a considerable amount of wastage shipping the LOX from plants to the missile units. Using fixed missile sites with their own LOX plants would substantially reduce LOX wastage and increase the daily number of missile launches possible.

Although a fixed site would be more efficient, the missile would have to be launched from locations well within the range of Allied medium bombers, and so any fixed site was likely to be heavily bombed. Such a site could be fortified,

The Walter catapult for the FZG-76 was eventually configured as a modular unit to make it easier to assemble in the field. This is a partial launcher preserved at the Eperlecques Museum near the Watten Bunker, which can be seen in the background. This launcher is missing the distinctive blast deflector found at the end of the catapult. (Author's collection)

In the mobile batteries, the A-4 missile was towed to the launch site on a Meillerwagen, which erected the missile on the launch pad prior to fueling. (MHI)

but this would add to the expense of the program. Two different bunker designs were prepared in 1942 including sketches and architectural models. The B.III-2a design envisioned erecting the missile inside the bunker and then towing the launch pad outside the bunker for launch; the B.III-2b design had two openings in the roof which would permit the missiles to be elevated from within the protective confines of the bunker and launched from the roof.

The alternative to fixed basing was mobile basing. This would require a mobile erector system to place the missile vertically on its launch pad, and it would require that all the elaborate testing and fueling equipment be re-packaged to fit on either railway cars or trucks and trailers to accompany the launcher into the field. While this launch configuration would be less vulnerable to air attack than a fixed site, it would be far less efficient and the rate of fire considerably less.

The head of the A-4 program, Oberst Walter Dornberger, laid out the various launch options in a study completed in March 1942. The study suggested that fixed sites could be created similar to the U-boat bunkers being built on the French Atlantic coast that would be impervious to aerial attack. However, army artillery officers favored a mobile basing system, as they were not convinced that any structure could withstand repeated air attacks and

Trials of air-launched FZG-76 cruise missiles took place in 1943, with the He-111H medium bomber finally selected as the most suitable carrier. This is one of the test launches at Peenemünde; the operational aircraft launched the FZG-76 from under the starboard wing. (NARA)

still remain functional. The issue wasn't simply the bunker itself, but the roads and railroad lines leading to the bunker, which would be needed to provide a supply of missiles and fuel. Although a final decision was put in abeyance until the A-4 missile proved viable, initial design work began on a mobile missile launcher in early 1942 including both road-mobile and rail-mobile options. Some more exotic launch options were considered, including a submersible launch barge that could be towed behind a U-boat. None of these progressed beyond paper designs.

The problem posed by the need for liquid oxygen for the A-4 led to the first construction effort connected with the V-weapons. In October 1942, a technical mission was sent to northern France and Belgium to inspect potential locations for the creation of two plants capable of producing 1,500 tons of liquid oxygen per month. The sites selected were Tilleur near Liège in Belgium, codenamed WL; and Stenay in the French Ardennes, codenamed WS.

The KNW launch bunker

The first successful launch of an A-4 Feuerteufel (fire-devil) missile from the Peenemünde test site in October 1942 led to a discussion about the program between Hitler and armaments minister Albert Speer on November 22, 1942. Hitler was shown models of the proposed launch bunkers as well as details of the proposed mobile launchers. He agreed to a production plan for the missile, but made clear his preference for the bunker launch sites in addition to the Army's preferred mobile launcher option. As a result, Speer met with Dornberger in Berlin on December 22, 1942, to lay out the program in more detail. Speer instructed that the bunkers be designed to the special fortification standard (Sonderbaustärke) with a 5m-thick steel-reinforced concrete ceiling and 3.5m-thick walls. Each bunker would contain enough missiles for three days of launches, totaling 108 missiles, along with sufficient fuel and liquid oxygen. Each bunker would be manned by 250 troops. Construction of the first bunker somewhere in the Boulogne area would begin as soon as possible and would be followed at the end of June 1943 with a second bunker on the Cotentin Peninsula in France opposite southern England. The Organization Todt (OT), the paramilitary construction group that had come under Speer's control following the death of Fritz Todt in an airplane accident in February 1942, would undertake construction. Supervision of the missile construction program was undertaken by director-general of OT, Xaver Dorsch, due to the high priority afforded the program by Hitler.

An experimental railroad launch system for the A-4 missile was developed and tested in 1944, but its use was rejected due to the vulnerability of the European railway network to Allied air attack. The Red Army later captured this equipment and used it in the late 1940s in its early ballistic missile program. (NARA)

ABOVE

The Walter catapult was operated by a gas-generator cart at its end, which generated a powerful pulse by combining T-stoff (hydrogen peroxide) and Z-stoff (sodium permanganate) fuel. This was fed into a tube running through the catapult rail, propelling a piston, seen in the foreground, which was attached underneath the FZG-76 missile. The Imperial War Museum, Duxford, has the only complete V-1 launch system including a full Walter catapult and the associated equipment. (Author's collection)

A KRAFTWERK NORDWEST, WATTEN (EPERLECQUES), FRANCE

This was the first large launch bunker built for the army's A-4 ballistic missiles and it is shown here in its planned configuration. Design of the bunker was headed by Werner Flos, chief engineer in the offices of Organization Todt-Zentrale in Berlin. This launch bunker was a further evolution from the B III design shown to Hitler in 1942. The most significant change was a decision in March 1943 to shift a planned liquid oxygen plant and its five compressors from Stenay to this bunker, leading to its enlargement. Codenamed Kraftwerk Nordwest (KNW: North-west Electrical Works), it was located near the Eperlecques woods, though the Germans usually called it the Watten Bunker due to its proximity to the nearest railroad station.

The design incorporated three principal elements: a railroad station, a missile assembly and preparation vault and the liquid oxygen plant. The north side of the bunker contained a major railroad station below it with a protected tunnel coming in underground on the west side through a fortified tunnel that connected to the main Calais–St. Omer line to the east of the site via a spur-line. The plan was to ship standardized trains to the KNW consisting of 20 R wagons; 10 X wagons and one personnel car. The R wagons each contained a single A-4 missile minus its "Elefant" warhead while the X wagons carried two Elefants each plus other associated missile components. On arriving at the KNW, the missiles were erected inside the vaults on the north side of the building and their warheads fitted. They were placed on a rail-mounted launch pad, fueled, and then moved in a fueled and erected state out of one of two armored doors on the south side of the building to a launch plaza. The walls of the exit tunnels were constructed in a chicane pattern to reduce the shock wave from the rocket exhaust during the missile launch, as there were no doors on the exterior exit. A large command tower was located in the center of the south side of the building where the launch operation was supervised. The north side of the building contained the liquid oxygen factory along with its five Heylandt compressors located in five vaults. These had the capacity to produce 50 tons of liquid oxygen per day, about enough for 20 missile launches. The bunker also had insulated storage tanks for liquid oxygen with supplies adequate for three days of launch operations. The original plan called for completion by December 31, 1943, in order to start missile launches against London by the end of the year. The building was roughly 92m wide and 28m high (300 × 90ft) requiring about 120,000m³ of concrete, roughly equivalent to that of 435 London Hilton hotels.

Allied intelligence became aware of the site in the summer of 1943 and, after consultation with civil engineers, decided to attack it after the first major pouring of concrete but before the concrete had the time to fully harden. The first major raid was conducted by 224 B-17 bombers of the US Eighth Air Force on the afternoon of August 27, 1943, delivering 366 tons of bombs of which 327 landed on the site, mainly the north side. Four more raids were conducted from August 30 to September 7, substantially destroying the north side of the building. The damage was so great that the Wehrmacht had to abandon use of the site for V-2 launches and instead salvaged what they could be completing parts of the south side of the building as an oxygen manufacturing plant dubbed Betonklotz (concrete block) to support missile operations. During early 1944, three oxygen compressors were installed at the site. After the start of Operation *Eisbär*, the remainder of the bunker was attacked by the RAF using Tallboy bombs, one of which penetrated the building and put an end to any further construction.

Kraftwerk Nordwest, Watten (Eperlecques), France (caption overleaf)

The enormous internal volume of the Kraftwerk Nordwest at Watten is evident from this photo inside one of its cavernous halls. The painting on the wall depicts a full-sized V-2 missile. (Author's collection)

The survey of potential launch sites began in the final days of December 1942. The team concentrated on sites in the Artois region of France and finally settled on a site near the town of Watten, since the area was easily accessible by rail and canal, there was a good local electrical power-grid, and there were several forested sites that appeared suitable for construction while at the same time being remote enough to prevent the local French villagers from observing the work. The first bunker was given the cover-name KNW (Kraftwerk Nordwest: Northwest Electrical Works). Initial plans were completed on February 12, 1943, and it was decided to merge the planned Stenay Oxygen Plant within the KNW Bunker. This meant enlarging the bunker considerably beyond that envisioned in the preliminary studies, requiring some 120,000m^3 of concrete and about 360,000 tons of material during the four months of construction. Besides the bunker itself, the plan included a substantial upgrade to the neighboring railroad lines to permit re-supply once the missile campaign began, and also included preliminary efforts to create supporting sites for stocking missiles and other necessary supplies. The nearby town of Wizernes was selected as the location for the main supply

This is the model of the B.III-2b bunker design shown to Hitler in November 1942, which initiated the A-4 bunker program. This version erected the missile inside the bunker then used an elevator to move the launch pad outside for launch. It was smaller than the eventual Watten Bunker, lacking a liquid oxygen plant. (MHI)

base, codenamed SNW (Schotterwerk Nordwest: Northwest Gravel Works). A limestone quarry in the town was selected since it would permit extensive tunnels to be dug for sheltering the missiles prior to delivery to the launch bunker. Hitler approved the plan on March 29, 1943, with the KNW Bunker scheduled to be ready for combat by December 31, 1943.

The Tausenfüßler supergun site

The second V-weapon to receive approval for construction was the HDP pump gun. Due to its enormous length of 127m, the gun could not be made mobile and it would inevitably have to be deployed from a fixed site with the tube inclined at a set angle. The most obvious solution was to place it in some form of underground fortification, whether natural or man-made. The task of finding a launch site was handed to Major Bock of Festung Pioneer-Stab 27, the

This shows a test version of the Tausenfüßler HDP supergun at the Wehrmacht's Hillersleben artillery test range after its capture in 1945. (MHI)

The Wiese Bunker at Mimoyecques has been converted into a museum, and a replica of a single Tausenfüßler has been created in one of the actual gun chambers. In reality, five guns were ganged together vertically in each drift. (Author's collection)

fortification regiment of LVII Corps, Fifteenth Army, based in the Dieppe area. A study in early 1943 concluded that a suitable hill with a rock core would be ideal as the gun tubes could be placed in drifts (inclined tunnels) and the support equipment and supplies located in tunnels adjacent to the firing tunnels.

B WIESE B711, MIMOYECQUES, FRANCE

The site for the Tausenfüßler supergun was constructed in a limestone hill about 5km north of the Hidrequent lime quarries, and codenamed Wiese (meadow) and Bauvorhaben 711 (B711: construction project 711). Work began in September 1943 by constructing rail lines to support the construction. Shafts were dug starting in October, which prompted two Allied air attacks on November 5 and 8 that delayed work for about a month. The original configuration of the site was for two tunnel complexes with a total of 50 gun tubes, but after the air raid, the western shaft was abandoned after little work had been completed. The remaining eastern complex consisted of five drifts angled at 50 degrees to the horizontal reaching 105m (340ft) below the hilltop. The five drifts exited the hilltop through a concrete slab 30m wide and 5.5m thick (100 × 18ft). Large steel plates protected the five openings and each port had a special armored door. The internal tunneling and elevator shafts were quite extensive to support the guns during operation, and had the site become operational, about 1,000 troops of Artillerie Abteilung 705 and other supporting units would have been deployed at Mimoyecques.

The HDP Tausenfüßler 15cm supergun was 127m (415ft) long. The distance from the breech to the first propellant charge was 6m, and then 3.2m for subsequent chambers. Each of the five drifts contained a stacked cluster of five gun tubes for a total of 25 gun tubes planned for the site. By the summer of 1944, only three of the five drifts had been completed.

Although there were a dozen attacks prior to D-Day, these were largely ineffective due to the durable configuration of the site. This changed on July 6, 1944, when the RAF struck again with 16 Lancasters carrying 6-ton Tallboy bombs. One Tallboy directly impacted the concrete slab on top of the complex, collapsing Drift IV. Three other Tallboy bombs penetrated the tunnel system below, creating extensive damage. Some effort was made to clean up the debris, but by late July it was obvious that the damage was too severe to justify continuation of the construction, especially since the RAF could very well stage additional Tallboy raids. Plans were made to reconstruct the Tausenfüßler gun battery at the Rixtent B81 liquid oxygen facility, but this never transpired. Although not formally abandoned, the Canadian 3rd Infantry Division overran the Mimoyecques site on September 5, 1944.

A site was selected at a limestone hill near Mimoyecques on the Pas-de-Calais. Codenamed Wiese (meadow) and Bauvorhaben 711 (construction project 711), initial construction work for support tunnels began in late May 1943 even though the gun concept had not yet been fully proven by full-scale tests. The initial configuration consisted of two gun complexes, each with five drifts 130m long, which could each accommodate five HDP barrels for a total of 50 guns. The work attracted the attention of the Royal Air Force (RAF) and three air raids were conducted against the site in early November 1943. In the wake of the air attacks, the Army decided to scale back the project by halting work on the western battery before any shafts were created, and concentrating on the eastern battery. The plans were to have the first cluster of five tubes ready by March 1944, and the full complex with 25 tubes by October 1, 1944. A full-scale trial at the Misdroy proving ground in April 1944 led to a failure after only 25 rounds had been fired. These problems led to a further reduction in the scope of the eastern

ABOVE
The Tausenfüßler fired the 15cm Sprenggranate 4481 projectile that weighed 97kg at the time of launch, seen here during technical evaluation at Aberdeen Proving Ground in Maryland after the war. (MHI)

RIGHT
Unlike a conventional cannon, the HDP supergun used multiple propellant chambers angled off the main chamber that were electrically detonated in sequence to propel the projectile. This is a test example at the Hillersleben proving ground. (MHI)

battery at the Wiese site from five drifts to three although work had already begun on some of the other drifts. Later tests in late May 1944 achieved a range of 90km, insufficient to reach London, but at least proving the feasibility of the concept. Artillerie Abteilung 705 was organized in January 1944 under Oberstleutnant Georg Borttscheller to operate the Wiese gun complex.

The V-1 waterworks

The FZG-76 cruise missile was the last of the V-weapons to receive deployment approval. The launch system for the missile was determined by the engine choice, a simple pulse-jet engine. Unlike gas-turbine jet engines, which can operate from a cold start, a pulse-jet engine requires a strong air-flow through the exhaust chamber before the engine can operate. The only way to accomplish this is to launch the missile using a secondary propulsion system, at which point the pulse-jet can be ignited. Fieseler looked at a variety of launch approaches for the FZG-76, usually based on long rail launchers since this provided sufficient time for the pulse-jet to ignite. The first approach was a Rheinmetall-Borsig rocket sled mounted behind the missile, but after trials this was rejected in favor of a steam catapult design from Hellmuth Walter Werke (HWW) called the WR 2.3 Schlitzrohrschleuder (split-tube catapult). A small gas generator trailer was attached to the base of the launch rail which mixed a combination of T-stoff (hydrogen peroxide) and Z-stoff (sodium permanganate) to create high-pressure steam that was pumped into a tube inside the launch rail box propelling a piston, connected underneath the missile. This system was somewhat similar to the steam catapults used on aircraft carriers except for the method of generating the steam. The Walter catapult was attractive since it provided more than enough power to get the missile airborne, and also was cheap to operate since it was reusable. On the negative side, it required the use of a very long 49m (160ft) launch rail, which was cumbersome to deploy. In the rush to deploy the FZG-76, other alternatives such as zero-launch solid rocket boosters were not seriously explored. The awkward launch rail would prove to be the Achilles heel of the V-1 missile system.

As was the case with the Army, so too was there a strenuous debate within the Luftwaffe over fixed versus mobile basing. The Luftwaffe's Flak arm was assigned responsibility for the launch sites, and the Flak commander, General der Luftwaffe Walther von Axthelm, wanted the missiles deployed in a large

Wasserwerk St. Pol near Siracourt was the only one of the first four waterworks to be nearly completed. This drawing from one of the wartime technical intelligence reports shows the state of the building in the summer of 1944. The precise launcher configuration has never been determined; Allied sketches show a single launch ramp as seen here but German accounts indicate that two launch rails were planned. (NARA)

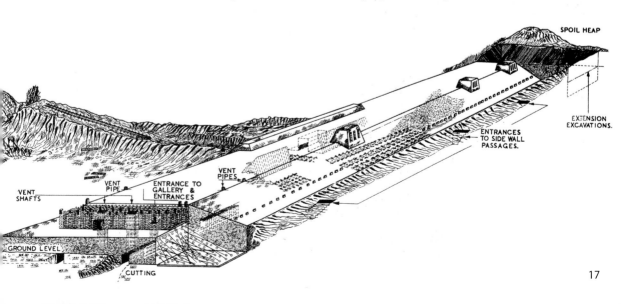

SPOIL HEAP

EXTENSION EXCAVATIONS.

ENTRANCES TO SIDE WALL PASSAGES.

VENT PIPES

VENT PIPE

ENTRANCE TO GALLERY & ENTRANCES

VENT SHAFTS

GROUND LEVEL

CUTTING

number of small "light" launch sites that could be easily camouflaged. However, the head of the Luftwaffe production program, General-feldmarschall Erhard Milch, knew that Hitler favored large launch bunkers, so he argued for this approach. A compromise was finally worked out during a meeting with the head of the Luftwaffe, Hermann Göring on June 18, 1943, with a plan to create four heavy *Wasserwerk* (waterworks) launch bunkers along with 96 light installations.

On the night of July 24, 1943, the RAF began Operation *Gomorrah* intended to destroy the city of Hamburg. The raids continued for several nights, starting a firestorm that devastated the city and left more than 50,000 dead. Hitler was

The bunker at Sottevast was intended to serve as a storage facility and base for a V-2 battalion near Cherbourg. It was only partially completed when the US Army overran the site in late June 1944, as seen here. (NARA)

infuriated, and critical of the Luftwaffe both for their failure to stop the attacks and their inability to retaliate. When Speer raised the issue of the FZG-76 missile program on July 28, 1943, Hitler enthusiastically approved the program, insisting that they be ready to pulverize London by the year's end in concert with the A-4 ballistic missile and the Tausenfüßler supergun. The complex of secret weapons sites in the Pas-de-Calais area were collectively called the *Sonderbauten*: "Special Construction" by Organization Todt.

The first of the heavy bunker sites were Wasserwerk Desvres located near Lottinghen and Wasserwerk St. Pol located near Siracourt, both in the Artois region of northeastern France. Another two would follow on the Cotentin Peninsula west of Normandy, Wasserwerk Valognes near Tamerville and Wasserwerk Cherbourg at Couville to the southeast of the port; the eventual goal was ten *Wasserwerke*. Even though this program started several months later than the A-4 ballistic missile program, the plan was to have the four *Wasserwerke* operational at the end of December 1943 to start the missile

C WASSERWERK NO. 1 ST. POL, SIRACOURT, FRANCE

The FZG-76 cruise missile bunker design was codenamed *Wasserwerk* (waterworks) by the Luftwaffe and four of the initial design were started at Siracourt and Lottinghen in the Pas-de-Calais and Tamerville and Couville on the Cotentin Peninsula. Of these, only Wasserwerk St. Pol near Siracourt came anywhere near completion due to Allied bombing of the sites.

This bunker design was based on the lessons of the destruction of the Watten Bunker, and used a new construction technique called *Verbunkerung*, which attempted to minimize the vulnerability of the bunker during construction. The structure was about 215m long, 36m wide and 10m high (700 × 120 × 30ft), a very low and long structure compared to the earlier Watten design and requiring some 55,000m³ of steel-reinforced concrete. The precise configuration of the finished bunker remains a bit of a mystery as none was ever completed, and full architectural plans have not been found. The design was basically an elongated tunnel with railroad access at the ends for supplying the bunker. Presumably, the interior would have been used for assembling and preparing the missiles for launch as well as housing the launch battery. There is some mystery about the intended

launcher configuration. The remains of the site at Siracourt suggests that a launch ramp would have emanated from the center of the building, and Allied intelligence presumed a single Walter catapult would have been fixed here. However, some German accounts have suggested that each site would have two launchers, so it is possible that the earthen launch ramp would have had two parallel launch ramps instead of only one seen here.

In spite of Allied bombing, Wasserwerk St. Pol at Siracourt continued construction until June 25, 1944, when the site was hit by 16 Lancaster bombers carrying the 6-ton Tallboy bomb. By this stage, about 90 percent of the concrete work was complete except for the sections at either end. However, the earth core had not yet been excavated from the insides of the structure. One Tallboy directly impacted the center of the roof, completely penetrating the structure, while another impacted within feet of one of the walls, causing significant damage. In total, Siracourt was subjected to 27 attacks with 5,000 tons of bombs. Even without the *coup de grace* of the Tallboy attack, the site was so badly torn up by the incessant bombing that it is hard to see how it ever could have been supported by rail transport.

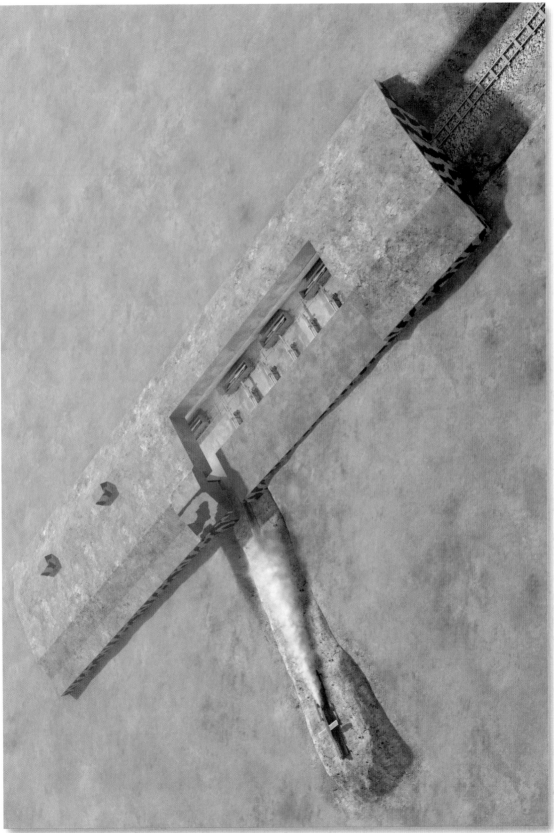

campaign against London, plus four additional bunkers by March 1944. However, work on the sites was badly delayed by other priorities as Organization Todt was stretched thin by its commitments to reinforcing the Atlantic Wall for the expected Allied invasion as well as a major rebuilding effort in Germany after the RAF's "Battle of the Ruhr" bombing campaign in the autumn of 1943.

THE ALLIES INTERVENE

British intelligence had some significant breakthroughs in discovering the German secret weapons programs in 1942. Resistance organizations in Poland and Luxembourg forwarded reports from forced laborers who had worked at the Peenemünde test center. After the RAF conducted reconnaissance flights, Churchill authorized Operation *Hydra*, which was carried out on the night of August 17/18, 1943. The bomber attack forced the Wehrmacht to abandon plans to mass-produce the A-4 missile at Peenemünde and to look for other alternatives. In conjunction with technical problems with the missile itself, the raid managed to push back the operational deployment of the A-4 missile by several months.

The discovery of the German test site also led the RAF to pay special attention to any unusual construction in France, especially in the Pas-de-Calais area closest to Britain. In May 1943, aerial reconnaissance first showed the start of construction of the large facility in the Eperlecques Woods near Watten. While the RAF was familiar with the numerous coastal bunkers that formed the Atlantic Wall program of coastal defenses, the Watten Bunker was extremely odd due to its distance from the sea. Although there were suggestions that it might be an operations control bunker, the details of the construction led to the growing conviction that it was part of a German secret weapons program. The US Army Air Force (USAAF) staged a daylight precision attack on August 27, and a second attack was staged on September 7, 1943.

The RAF raid on Peenemünde forced the Wehrmacht to abandon plans to mass-produce the V-2 missile there. Instead, the main center became the underground Mittelwerke tunnel complex in the Harz Mountains near Nordhausen. (NARA)

The Mittelwerke complex was so well hidden that the Allies did not bomb it. This shows one of the camouflaged entrances after its capture by the US Army in April 1945. (NARA)

The heavy damage sustained by the KNW Bunker forced a significant revision in German plans for the missile sites in September 1943. The partially completed south side of the bunker had collapsed and so it could no longer be used as intended as a missile launch site. The northern half of the bunker could be repaired, so its role was downgraded to that of a liquid oxygen factory. The SNW supply depot in the limestone quarry at Wizernes suddenly attracted more attention since it held the potential to be converted quickly into a ballistic missile launch base. Instead of constructing a bunker in the open, a reinforced concrete dome could be created above the launch facility, and then the cavity for the launch bunker could be carved into the quarry, limiting the amount of damage that could be inflicted on the structure during construction.

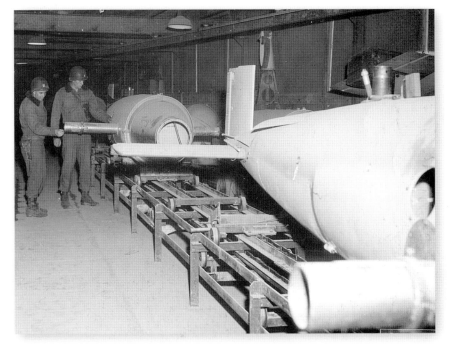

As in the case of the V-2, Allied air attacks on exposed V-1 production plants led to a shift to the Mittelwerke for mass-production late in 1944 with the Mittelwerke II "Zinnstein" facility taking over the majority of production after November 1944. (NARA)

The heavy damage inflicted on the Watten Bunker forced the Wehrmacht to abandon plans to use it for launching V-2 missiles. Instead, the undamaged southern side of the building was completed for use as a liquid oxygen production bunker. This shows the preserved bunker from its southern side. Had it been completed as a missile bunker, large portals for the V-2 would have been exited on this side to launch the missiles outside the bunker. (Author's collection)

The damage inflicted on the Watten Bunker led to a revised construction method for the *Wasserwerk* bunkers, called *Verbunkerung* or *Erdschalung*. The construction began with (1) a simple set of trenches; (2) the basic side wall foundation; (3) excavation for the complete side walls; (4) completion of the side wall foundation; (5) construction of roof over an earth core; and (6) excavation of the earth core, completing the construction. (NARA)

The Watten debacle turned the Germans' attention to mobile basing. The November 1943 plan envisioned a total of four A-4 battalions along the French coast, each with three batteries. Two battalions would be stationed in Artois near the Pas-de-Calais with 26 launch bases and five guidance bunkers; one battalion with six launch bases and two guidance stations would be completed near Dieppe, and one battalion with nine launch bases and three guidance bunkers near Cherbourg. Two other large bunkers associated with the A-4 ballistic missile were also added, the Reservelager West (RLW: Reserve Store West) near Brix/Sottevast and the Ölkeller Cherbourg (Cherbourg Oil Cellar) near Brécourt. As late as November 1943, the precise role of these two facilities had not been finalized. Sottevast was a large protected bunker, comparable in size to the original Watten design, which could be used for housing one of the mobile missile regiments as well as serve as a store for about 300 missiles, fuel and other supplies. Brécourt included a series of protected tunnels for storing the missiles.

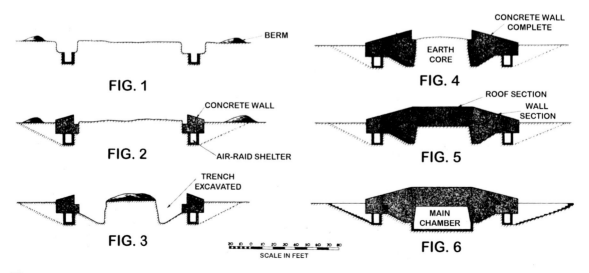

The Kraftwerk Nordwest missile bunker at Watten was hit by USAAF heavy bombers starting on August 27, 1943, which wrecked the northern side of the building. This photo was taken in the summer of 1944 after the site was revisited by RAF Lancasters armed with Tallboy bombs, which account for the enormous crater in the foreground. (NARA)

The first Allied air attacks forced changes in the construction plans for the *Wasserwerke* as well. Instead of creating a bunker in the traditional fashion, a new six-step process called *Verbunkerung* or *Erdschalung* would be undertaken to shield the bunker from air attack during construction: the side walls would be created first while protected by an earthen berm, the cavity between the walls filled, the roof poured over a temporary inner earthen core, and then the internal cavity excavated to form the main bunker chamber. Completion of the Wasserwerk 1 design permitted construction to begin in late September 1943. The bunker was essentially a 212m long protected tunnel. Each *Wasserwerk* could contain up to 150 FZG-76 cruise missiles along with their associated fuel and support equipment, and this entire inventory could be expended in one or two days depending on the launcher configuration. Although the design was ready, the construction program for the *Wasserwerke* came at an inopportune time and the deadline for starting the

With the Watten Bunker no longer viable as a missile launch base, the tunnel complex in the chalk quarries at Wizernes was expanded to include a massive missile complex under a special reinforced concrete dome. Two tunnels would have exited the complex to permit V-2 missiles to be launched from the open plaza below the dome that were sealed by Allied bombing. The tunnel entrance to today's museum as seen here is based on the Ida railroad tunnel. (Author's collection)

missile campaign in December 1943 was not met. Organization Todt was diverted to reconstruction efforts in Germany as a result of RAF attacks on the Ruhr. The *Sonderbauten* effort required about 1.83 million cubic meters of concrete, competing with other construction programs such as the Atlantic Wall coastal defense program. To further undermine the program, there were significant delays in the mass-production of the missiles after the RAF had bombed the Fieseler plant in Kassel on the night of October 22/23, 1943.

Graveyard for the RAF: Stellungsystem-I

The last of the V-weapons sites to enter construction were the FZG-76 cruise missile "light" launch sites. Although Generalfeldmarschall Milch, the head of Luftwaffe production, originally favored the heavy waterworks, by September 1943 he had changed his mind. He became convinced that a large number of FZG-76 missile sites would inevitably divert the attention of the RAF away from Germany, relieving the pressure on the Luftwaffe, and making the bombers more vulnerable to fighter attack since precision daylight missions would be needed. He dubbed the *Sonderbauten* effort on the Pas-de-Calais "the graveyard of the RAF." In mid-August 1943, Luftwaffe personnel began to visit locations in the Pas-de-Calais area, informing the local farmers that their property was being requisitioned. In many cases, the farmers were allowed to continue to work their farms, though they were kept away from some areas that were being used for construction. This was the first step in the creation of Stellungsystem-I, the "light" launch sites for the FZG-76 missile. The plan was to deploy the newly formed Flak Regiment 155(W) in these sites. At the time, FR 155(W) was still training on test launchers at Peenemünde, and it was anticipated that it would have four launch battalions ready by the start of the missile campaign scheduled for December 1943. Each battalion had four launcher batteries with four launchers each, for a total of 64 launch sites under the regiment's control. These 64 launch sites formed Stellungsystem-I and they were located primarily in the Pas-de-Calais region of northeastern France from Lille through Dieppe. Besides the 64 primary launch sites, steps were also

This wartime intelligence drawing shows the typical layout for a Bois-Carré site. The main facilities include:

(1) the launch rail and protective wall;
(2) fire-control bunker;
(3) three "ski" stowage buildings;
(4) the compass correction building;
(5) personnel bunker;
(6) missile servicing building;
(7) launcher service building;
(8) water reservoir;
(9) cistern and pump station;
(10) storage building.

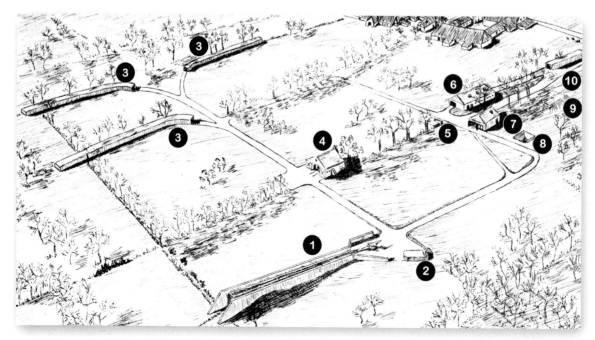

underway to create Stellungsystem-II, with a further 32 sites that were intended to serve as reserve launch locations as well as supply bases, two per launch battery. The last portion to be started was Stellungsystem-III, located southwest of the Seine from Rouen to the Cotentin Peninsula in lower Normandy. Stellungsystem-III was located for attacks on British cities in southern England after London had been destroyed and would be manned by a second regiment, FR 255(W), which was organized in the late spring of 1944.

The Bois-Carré sites protected the Walther catapult with reinforced walls on either side. This is a typical example near the Chateau de Sorrelerie in the Cherbourg area, one of the batteries of Abteilung IV. (NARA)

Construction of Stellungsystem-I began in the late summer and early autumn of 1943 under the direction of the Luftwaffe's Sonder Pionier-Stab Frisch (Special Engineer Staff Frisch) of the Fifteenth Army (AOK 15) in the Pas-de-Calais area and Sonder Pionier-Stab Beger of AOK 7 in the Normandy area. These units supervised the construction work undertaken by Organization Todt, though most of the actual labor was subcontracted to French construction firms. Each site centered around a platform for a Walter catapult, protected on either side by a concrete blast wall. Each site included a standard assortment of support buildings, though the layout of the buildings varied from site to site. The layout was intended to facilitate a high missile launch rate. The Walter launch rail could be fired and reloaded in 20-minute intervals so it could launch 72 missiles per day at its maximum rate. This was not entirely realistic since each site had accommodation for only 21 missiles, so a more realistic rate of fire per site was about 20 per day.

Certain of the structures essential for the launch process were located near the launch rail, while other preparation buildings were arranged based on the layout of the terrain, for example placing the long storage building along hedgerows to provide natural camouflage. The missiles arrived at the site, usually by truck, in a partially assembled form on special dollies. The launch battery had to complete the final assembly of the missile, as well as fuel, fuze,

Designation	length × width × height (m)	Description
Abschussrampe	58.4 × 9 × 5	Launch ramp
Kommandostand	5.9 × 3.6 × 2.7	Launch bunker
Eingangslager	29.5 × 4.3 × 3.0	Ready storage building
Montagehalle	21.4 × 8.25 × 4.05	Preliminary assembly building
Werkstat/maschinenhaus	14 × 8.1 × 3	Workshop
Wasserbehalter	10.1 × 10.1 × 3	Water reservoir (200m³)
Stofflager	7.8 × 6.2 × 3	Propellant storehouse
Betriebswasser zistern	15.3 × 6.8 × 3.8	Water cistern
Vorratsblager	82 × 4.3 × 3.35	Main "ski" storage building
Richthaus	14.8 × 17.6 × 9.7	Non-magnetic guidance adjustment building
Zünderbunker	5 × 4 × 2.8	Fuze storage bunker
Unterstand	14.2 × 11.6 × 3.3	Personnel accommodation
Unterstand für trafo	5.5 × 4.8 × 3.7	Transformer shed
Pumpstation	8.8 × 6.6 × 5.6	Pump station

The rocketgun coast: German *V*-Weapon sites in France 1944

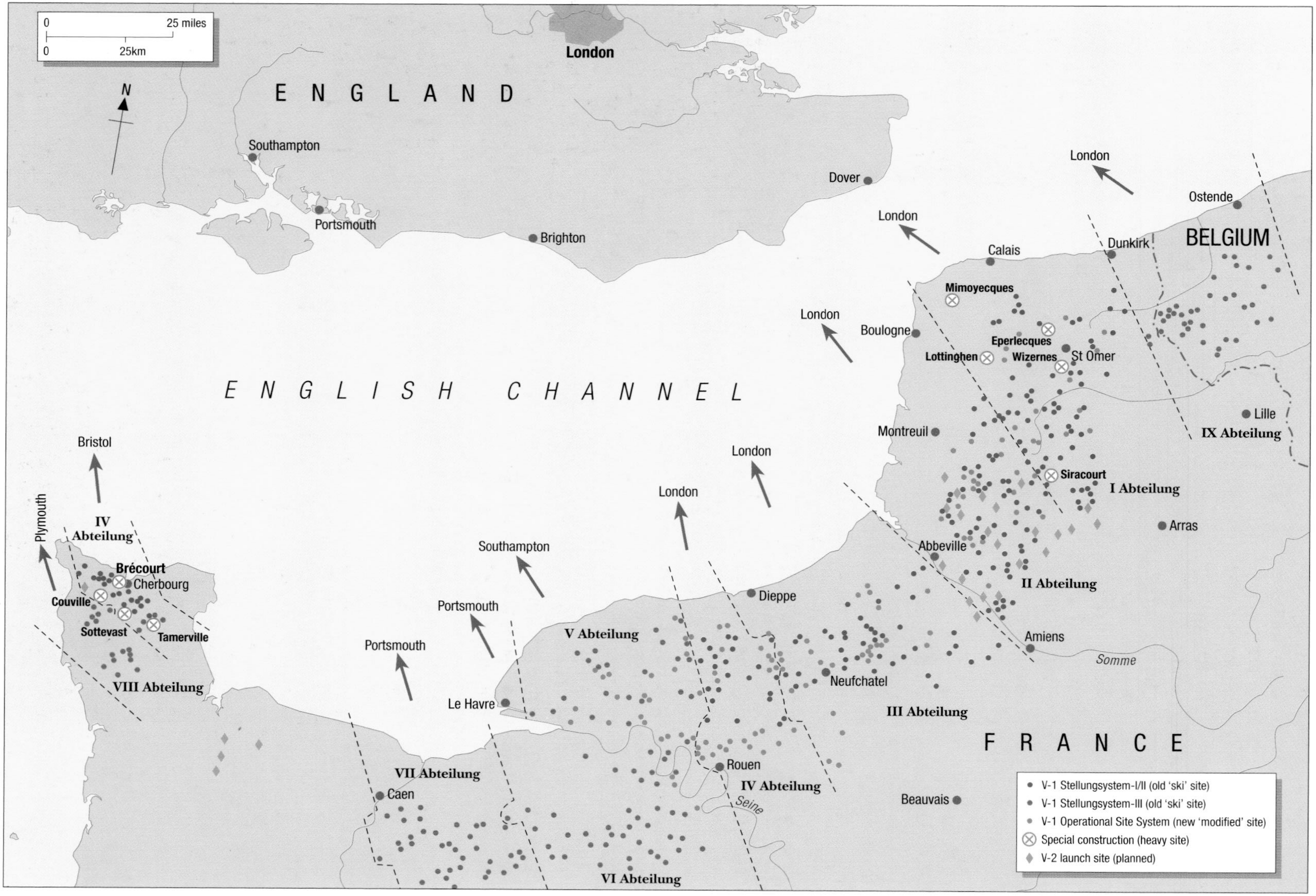

Legend:
- ● V-1 Stellungssystem-I/II (old 'ski' site)
- ● V-1 Stellungssystem-III (old 'ski' site)
- ● V-1 Operational Site System (new 'modified' site)
- ⊗ Special construction (heavy site)
- ◆ V-2 launch site (planned)

ENGLAND

London

Southampton
Portsmouth
Brighton
Dover

ENGLISH CHANNEL

N

25 miles
25km

London
London
London
London
London
Southampton
Portsmouth
Portsmouth

BELGIUM

Ostende
Dunkirk
Lille
IX Abteilung
Arras
Calais
St Omer
Siracourt
I Abteilung
Mimoyecques
Eperlecques
Wizernes
Lottinghen
II Abteilung
Amiens
Boulogne
Montreuil
Abbeville
Somme

FRANCE

III Abteilung
Neufchatel
Beauvais
Dieppe
IV Abteilung
Rouen
Seine
V Abteilung
VI Abteilung
Le Havre
VII Abteilung
Caen

Cherbourg
Brécourt
Tamerville
IV Abteilung
Sottevast
VIII Abteilung
Couville
Bristol
Plymouth

and arm the missile. Besides the missile-related facilities, most sites also had a few 20mm or 37mm Flak guns to protect the site from air attack.

The primary construction material for Stellungsystem-I/-II was concrete bricks (cinder-block), consuming some 336,000m³ of concrete and 180,000 concrete bricks. With the exception of the command bunker located near the launch rail, none of the structures was fortified. The principal building types are listed on page 25 by their German designations and role.

While construction was underway at the launch sites, a parallel program was underway to create storage facilities for the missiles in the neighboring areas. In some cases these were converted from tunnels or other existing structures. The first nine of these were planned for the Pas-de-Calais while the last four were located in Normandy.

As construction of Stellungsystem-I progressed in the autumn of 1943, it increasingly came to the attention of British intelligence. French resistance organizations had noted the flurry of activity in the Pas-de-Calais region and began to systematically collect information on the sites which was forwarded to London. At first, the role of the sites was not very clear as they were not especially large, and British intelligence had a very cloudy view of what types of weapons the Germans were developing. The Joint Intelligence Committee (JIC) was aware of the possibility of some form of jet-propelled missile and in late August 1943, a FZG-76 test fired from Peenemünde crashed on the island of Bornholm and Danish resistance spirited out photos of the wreckage. By November, the scattered evidence began to coalesce, especially after RAF reconnaissance aircraft again photographed the test site at Peenemünde. Sitting on a long rail was a small aircraft, and nearby were storage buildings suspiciously similar to those being erected in France. Photographic reconnaissance missions over France were accelerated, and a site near Bois-Carré seemed to be nearest to completion. The most distinctive element of these sites was catapult walls ominously pointed towards London, and several "J"-shaped

FZG-76 missile supply depots

Location	Codename	Missile capacity
Ytres Canal Tunnel	Karl-Theodor	1,000–1,500
St. Leu-d'Esserent	Leopold	1,600–2,400
Nucourt	Nordpol	1,000–1,500
Bessancourt	Bertha	1,000–1,500
St. Maximin	Martha	1,000
Fort VII-Antwerp	Anton	240
Fort Cruybeke	Christa	200
Fort Rochambeau	Robert	150
Fort Hirson	Hildegard	150
Balleau	Biber	275
Mamers	Murmeltier	375
Luché	Luchs	300
Cherbourg	Chamäleon	300

The Bois-Carré sites used a large concrete structure designed to be non-magnetic. It was aligned on London and used to adjust the magnetic compass in the V-1 flight control system prior to the missile launch. This type of structure is easily identified by its distinctive arched entrance and this partially destroyed example is preserved at the Ardouval/Val-Ygot site. (Author's collection)

The distinctive "ski" buildings used for missile storage gave the Bois-Carré sites their name. When possible, they were buried for further protection, and protective walls erected on either side of the entrance, as is seen at this site of Abteilung IV south of Cherbourg near Martinvast. (NARA)

buildings, which were dubbed "ski" buildings by the photo interpreters. As a result, the Stellungsystem-I sites were nicknamed "ski sites" or "Bois-Carré" sites by British intelligence. By late November 1943, some 75 sites had been spotted in the Pas-de-Calais area, and seven near Cherbourg. In mid-November, a sub-committee of the JIC was created codenamed "Crossbow" to coordinate the intelligence collection directed against the German missile program. As a result, the German missile sites in general were often referred to as "Crossbow sites," a term that will be used here for convenience.

On December 1, 1943, the Wehrmacht operations staff created the 65.Armee Korps zur besonderen Verwendung (65th Army Corps for Special Employment), to command the planned missile attacks on London that were scheduled to begin at the end of the month. This corps headquarters was an unusual organization including both Army and Luftwaffe missile units, as well as the Tausenfüßler gun regiment and the various support units associated with the new weapons. As it transpired, the start of the missile campaign was delayed by production and technical problems with the missiles, but the corps set up its headquarters in St. Germain, France, in early 1944.

This Abteilung IV ski site near Cherbourg has extensive camouflage netting over the work area. The building is the non-magnetic compass alignment garage so typical of the Bois-Carré sites. (NARA)

The threat of air attack led FR 155(W) to make extensive use of camouflage nets around the launch sites. This is a "ski" stowage building at one of the Abteilung IV sites near Cherbourg. (NARA)

THE SITES AT WAR

Attacking the rocket gun coast

The first Allied air attacks on the new Crossbow sites began on December 5, 1943, when B-26 aircraft of the USAAF Ninth Air Force attacked three ski sites near Ligescourt. Due to weather conditions the results were poor, and the RAF became convinced that heavy bombers would be needed. The first Bomber Command attack took place on the night of December 16/17, 1943, on sites near Abbeville. The results again were poor due to the difficulties of conducting precision bombing against such small and bomb-resistant targets at night. The sites were difficult to damage as the buildings were fairly small and quite scattered. On the other hand, practically none of the buildings were fortified, so even near misses could collapse them. On December 15, the Joint Chiefs of Staff decided to begin employing US heavy bombers in daylight in saturation attacks, and when clear weather arrived on Christmas Eve, 672 USAAF B-17 and B-24 bombers delivered 1,472 tons of bombs on 24 ski sites. By the end of the year, 52 sites had been attacked and nine were believed to have been seriously damaged. Actually, seven sites had been put out of action of which three had been completely obliterated.

The leadership of the US Army Air Force became concerned about the growing diversion of heavy bombers from their primary mission, Operation *Pointblank*, the strategic air campaign against German industry. British intelligence had not been forthcoming about the Crossbow threat, and it was unclear why heavy bombers were needed when there was an ample supply of medium bombers and fighter-bombers in England that could be used without jeopardizing *Pointblank*. On December 29, 1943, the US War Department set up its own US Crossbow Committee under General Stephen Henry of the New Developments Division to study the V-weapons threat. Its initial efforts were directed at getting information from British intelligence sources, and in January 1944 it reported back that a pre-emptive strike against the missile sites was the best option. However, there still was doubt that US heavy bombers were the best suited to attack such sites. The USAAF Proving Ground Command at Eglin Field in Florida studied the issue and in February and March 1944, they began erecting mock Crossbow sites at the proving ground that were then attacked using a variety of aircraft types with various types of weapons from different altitudes. The tests convinced the US Crossbow Committee that the most effective tactics were treetop raids by fighter-bombers using delayed-action 1,000lb and 2,000lb bombs. Although briefed on US tests, Air Marshal Leigh-Mallory, head of the Allied Expeditionary Air Force, dismissed the US proposal to shift to low-altitude attack, arguing that the US tests were idealized experiments and that in real world conditions the low-flying aircraft would be at greater risk than the US trials suggested. This debate became embroiled in an even more contentious argument between the US air commanders and Leigh-Mallory over the objectives of the strategic bomber offensive.

The new-pattern sites

The December attacks were only the beginning of a long air campaign against the ski sites. Hitler's plan to start the missile campaign on London by the end of December 1943 proved impossible due to lingering technical and manufacturing difficulties with the new missiles. By the end of December 1943, the Luftwaffe was coming to recognize a supply of FZG-76 missiles probably wouldn't be available until March 1944 at the earliest, and therefore the missile launch sites would have to endure months of concentrated bombing. Oberst Wachtel, commander of FR 155(W), asked for heavy flak batteries to protect the sites against high-altitude attack by heavy bombers, but given the scattered nature of the *Stellungsysteme*, the flak arm was unwilling to weaken the air defenses of the Reich by transferring precious flak batteries. By January 1944, it was becoming apparent the existing *Stellungsysteme* would be too damaged to be functional when mass production of the FZG-76 finally began. The production of the Walter catapult did not begin until January 1944, and sufficient launchers were not completed until the end of February 1944. The *Stellungsysteme* were unredeemable; the stereotyped configuration of the sites made them obvious to Allied air reconnaissance, and their construction by French workers led to their early discovery through espionage. The air attacks terrified the French construction crews who ran away at the first sounds of aircraft; by the first week of January 1944, only two percent of the original workforce was available due to the mass desertions.

The new sites were almost invisible from the air until the Walther catapult was erected. This shows one of the launch sites of Abteilung IV on the Cotentin Peninsula abandoned after D-Day. The concrete launch pad and rails are camouflaged with hay. (NARA)

As an alternative, in January 1944, the first steps began to devise a new launch system. FR 155(W) had already begun to consider the needs of a minimal launch site without the distinctive buildings such as the "ski" buildings that disclosed the site's location. Wachtel argued forcefully that any new site system should be constructed solely by German military construction units to reduce the vulnerability of the sites to enemy espionage. The development of the new site system was entrusted to Oberst Schmalschläger. The Luftwaffe

decided to continue construction and repairs on the *Stellungsysteme* using available French construction firms in order to distract Allied attention from the new sites.

FR 155(W) began to man Stellungsystem-I and -II in hopes of starting a missile campaign sometime in March; the first nine batteries were in position on February 1, 1944. Allied intelligence assessments were alarmist, and the Crossbow committee estimated that the Luftwaffe would be able to start missile attacks at a rate of 215 to 770 missiles in an eight-hour period by late February, increasing to a rate of 430 to 1,640 missiles by late March. The estimates were wildly off the mark and an FR 155(W) command post exercise on March 2, 1944, assumed that between 770 and 960 missiles could be launched assuming the availability of all 64 launchers. However, the sites at the time had no missiles or catapults, and half of them had been damaged to the point of being unusable. The panicky intelligence assessments led to even heavier raids on the Crossbow sites, and on February 13, the Combined Chiefs of Staff gave the Crossbow mission priority over all other Allied bombing targets except for the German fighter industry. In March, a total of 4,250 tons of bombs were dropped in 2,800 sorties. The relentless Allied air attacks systematically pulverized the *Stellungsysteme*. According to the FR 155(W) regimental war diary, by the end of March 1944, nine sites had been destroyed, 35 seriously damaged and 29 had suffered medium damage.

In spite of the extensive damage to the sites, Allied intelligence concluded that the Crossbow attacks had failed to end the threat and on April 18, 1944, the secretary of the British War Cabinet, Sir Hastings Ismay, pressed Eisenhower for even more attacks. As a result, the next day the Crossbow campaign was given top priority even over the attacks on the Luftwaffe fighter factories, much to the dismay of the senior leadership of the USAAF. Since the attacks were to be carried out in daylight, the majority of the attacks were assigned to the US Eighth Air Force rather than the RAF Bomber Command. Both US and British medium bombers took part in the raids as well. During April 1944, the Crossbow attacks totaled some 7,500 tons of bombs in 4,150 sorties and the FR 155(W) war diary recorded that, by the end of the month, 18 sites had been destroyed and 48 suffered heavy damage.

Efforts by General Arnold to get the RAF to re-examine the Eglin report were ignored and in early May, the 365th Fighter Group was ordered to conduct a trial attack at low altitude against four ski sites using a single P-47 fighter per site, each carrying two 1,000lb delayed-action semi-armor-piercing bombs. The attacks inflicted Category A damage, enough to neutralize the site for several months, on three of the four sites with no US losses. In contrast, US heavy bombers required on average some 227 tons of bomb per site and 131 sorties to inflict comparable damage; while medium bombers required 231 sorties and 189 tons of bombs, a tremendous waste of resources. The RAF admitted that its own Mosquito light bombers had proven by far to be the most effective means to attack the sites, averaging 62 sorties and 40 tons per ski site to inflict Category A damage. The RAF would later admit that the brute force approach was little more than "sledgehammers for tintacks."

With no sign of any missile attacks, and with Allied intelligence now conceding that most sites would require two months of work to become operational, the high priority afforded the Crossbow campaign was finally rescinded in early May in favor of Operation *Pointblank* and air attacks in preparation for the Operation *Overlord* landings in Normandy. By early May, 24 of the launch sites had been destroyed and 58 had suffered serious damage according to the regimental war diary. By the time of the D-Day invasion, of the 96 launch sites completed, 83 were damaged beyond use and only two

One of the few essential elements of the new-pattern sites was this concrete pad that served as the base for the Walter catapult. The rails were used for an overhead crane to install a launcher while the small rectangular depressions were used to anchor a track for the steam-generator cart. This was one of the eight Abteilung IV sites located in the Cotentin area, possibly FSt.227 at Saint Colombé, which was abandoned due to the D-Day invasion. (NARA)

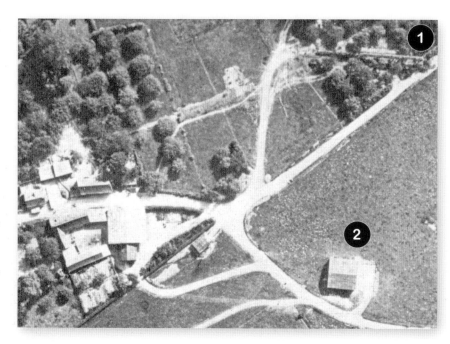

Even when the new sites were activated with their equipment, they were very difficult to spot from the air as can be seen from this wartime aerial reconnaissance photo. The only clear indications of the site are (1) the Walter catapult and (2) the compass correction building. (NARA)

would ever see combat use. However, the German ruse had worked and the efforts to continue repair work at the ski sites had enticed the Allied air forces to waste over 11,550 sorties and 16,500 tons of bombs from February to June in Crossbow missions against sites the Germans never planned to use.

In the spring of 1944 with the Crossbow campaign well underway, Oberst Schmalschläger's team had developed a new simplified site system. The firing sites were configured with an absolute minimum of permanent structures. Basic pilings for the launch ramp, a flat platform for the steam generator trolley, and a foundation for the non-magnetic guidance shed were made from concrete. The new sites were generally positioned near French farms where the existing buildings could be used for crew accommodation and storage. Certain of the specialized buildings such as the navigation correction building used prefabricated wooden sheds instead of concrete structures. The distinctive ski buildings were not used and missiles were either stored in available buildings or left under camouflage nets. When time permitted, some small structures were built, especially the steam generator preparation shed, workshops for preparing the missile, fuel storage sheds, and the launch bunker near the catapult, and in some cases, prefabricated structures were used. It took a work party of 40 men only about two weeks to construct such a site.

None of these buildings were especially conspicuous, and the new sites proved to be almost invisible to air detection until the launch ramps began to be erected in June 1944. To prevent their identification by the French resistance, the construction was undertaken solely by German military units, the Luftwaffe's Bau Pioneer Battalion Luftgau Belgien-Nord Frankreich (Belgium-North France Air Command Engineer Construction Battalion), and the Army's Sonder Pioneer-Stab Frisch (AOK 15) in the Pas-de-Calais and Sonder Pioneer-Stab Berger (AOK 7) in Normandy. Since these units did not have enough troops to carry out the work, they employed convict labor for much of the

Crossbow bombing campaign August 1943–March 1945		
Unit	Sorties	Tons of bombs
US Eighth AF	17,211	30,350
RAF Bomber Command	19,584	72,141
US Tactical Air Forces	27,491	18,654
RAF Fighter Command	4,627	988
Total	68,913	122,133

One of the buildings in common between the old and new pattern sites was the small fire-control bunker positioned near the end of the launch ramp. This example is from the old-pattern site at Ardouval/Val-Ygot and is semi-submerged with the view slit near ground level. (Author's collection)

When resources were available, the new sites sometimes included a few of the smaller buildings found in the original sites, such as this small bunker used to store the V-1 warhead fuzes. Usually, these bunkers were partially buried to protect the delicate explosives, but this example near Cherbourg was not completed before its capture by the US Army in late June 1944. (NARA)

construction on the assumption that the prisoners' contact with the outside could be restricted. The Walter catapult ramp took about seven to eight days to erect, and were only brought to the site at the start of the missile campaign.

A network of local caves, tunnels, and mines was taken over for use as improvised ordnance storage areas. In total, the "Operational Site System" consisted of five launch sites (*Feuerstellungen*) for each launch battery plus a support site, for a total of 80 launch sites and 16 support sites, located from Calais westward into lower Normandy. The original "ski sites" were then called *Stellungen alter Bauart* (old-pattern sites) while the new simplified sites were called *Einsatz Stellungen* (special sites).

Through signals intercepts, British intelligence was aware that the Germans were planning to deploy less conspicuous sites and such sites were mistakenly identified as early as February 1944. This was not the case as construction began only late in the month. The confusion was due to the large variety of support sites associated with Stellungsystem-I and -II. Besides the launch sites, there were the supply sites (*Versorgunungs Stellungen*); reserve sites (*Ersatz Stellungen*), as well as battery command sites (*Batterie-Befehls-Stellen*). The first new configuration site was found on April 26, 1944, near Belhamelin in the Cherbourg area, and so these sites were dubbed Belhamelin or "modified" sites. Allied intelligence estimated that by June 12, when the missile campaign began, 66 of the modified sites had been identified by a renewed reconnaissance effort in May 1944. However, the FR 155(W) war diary insisted that as late as

The Wasserwerk St. Pol at Siracourt was the most heavily attacked of the heavy sites. Although the concrete work was largely complete, as is evident from this Allied intelligence photo from the summer of 1944, the excavation of the interior earth core had not progressed very far before the bombing halted final construction. (NARA)

May 26, 1944, none had been located by Allied intelligence, presumably due to the lack of air attacks on these locations.

Although many subsequent studies later concluded that the initial Crossbow bombing campaign delayed the start of the missile offensive by six months, this was at best partly true. The strikes against the ski sites had no effect in delaying the start of the campaign, and indeed the German Flak officers later argued that the initial December 1943–January 1944 attacks on the first ski sites had the positive consequence of forcing the Luftwaffe to design more survivable sites months before the actual missile campaign could begin. The Crossbow campaign did largely derail efforts to deploy the FZG-76 from the *Wasserwerke*. The main delay in the start of the missile campaign was not the attacks on the launch sites, but rather the strategic bombing raids against the German aircraft industry, which delayed mass production of the FZG-76 until the spring of 1944 instead of the autumn of 1943 as planned. As a result, FR 155(W) did not have an adequate inventory of missiles, launch ramps, and other necessary equipment until late May 1944.

The Allied success in preemptive bombing of the first set of *Wasserwerke* forced the Organization Todt to try less conspicuous approaches to heavy launch sites. The extensive tunnel network created originally for V-2 storage at Brécourt near Cherbourg was converted to a V-1 launch site with this massive set of blast walls erected at the end of one of the tunnels to accommodate a Walter catapult. In the event, the site was captured before any missiles could be fired. (NARA)

Sonderbauten sites

The smaller number of *Sonderbauten* sites were also attacked as they were identified, starting in October 1943. Even though the Wehrmacht was increasingly skeptical of the value of these bases, Organization Todt continued their construction as a matter of pride to prove they could build in the face of bombing attacks. The proposed *Wasserwerk* sites fared very poorly under heavy bombing, in spite of the innovative construction techniques. Wasserwerk Cherbourg B8 near Couville in Normandy was heavily damaged by air attack on November 11, 1943, during the preparatory stages of construction and did not progress beyond excavation work due to repeated attacks through January 1944. Wasserwerk Valognes B7 at Tamerville, also on the Cotentin Peninsula, suffered a somewhat different fate. Work was delayed due to a general reconsideration of the prospects for the *Wasserwerke* after the November 11 raid. In the meantime, Erwin Rommel, assigned to reinvigorate the Atlantic Wall defenses in anticipation of the expected Allied invasion, stumbled on to the site while on one of his fact-finding tours of the Cotentin Peninsula. The launch site was very close to the command post of the 709th Division defending this sector, and so the Army commanders pressured Organization Todt to halt construction and divert the resources to the coastal defenses instead. As a result, the Tamerville site was downgraded to a reserve site with little substantial construction work. The *Wasserwerke* in the Pas-de-Calais region remained under construction by Organization Todt in spite of air attacks. Wasserwerk Desvres at Lottinghen was hit for the first time on February 24, 1944, eventually absorbing nine raids and 605 tons of bombs. The attacks undermined the side walls before the roof could be constructed, so in April 1944, Organization Todt was forced to abandon construction at the site. Wasserwerk St. Pol near Siracourt was first hit on January 31, 1944, but this

Wasserwerk No. 2 (Ersatz B8), Brécourt, France

D WASSERWERK NO. 2 (ERSATZ B8), BRÉCOURT, FRANCE

The size of the first four *Wasserwerk* V-1 sites made them dangerously conspicuous to Allied air intelligence, and so alternatives were developed in the spring of 1944. The German Army had taken over and expanded some French Navy tunnels near Brécourt in the western suburbs of Cherbourg, which were originally codenamed Ölkeller Cherbourg. These were intended to store A-4 ballistic missiles but when that program was delayed by technical difficulties, the Luftwaffe took over the site. It was variously codenamed Minenlager (mine storage) or Ersatz B8 since it was intended to replace the ill-fated B8 Wasserwerk at nearby Couville. Brécourt was also called Wasserwerk No. 2, as it was planned to use this as the model for a new series of protected launch bunkers. In March 1944, a second site of this type was considered for the Roche de Tronquet tunnels near Nardouet. These tunnels were in use by the Kriegsmarine, but would be modified with four sheltered launchers emanating

from the tunnels. It did not progress beyond the design stage.

Aside from adapting the tunnels, the main work at Brécourt focused on the creation of a pair of heavily protected launch areas, both of which would be fitted with a standard Walter catapult launcher. The neighboring tunnel complex could store 300 FZG-76 missiles, enough for about six days of launches. Due to its modest size and location near the coast, the role of the site escaped Allied attention as it was categorized off as just another part of the vast Atlantikwall coastal defense program. Most of the work on the western launcher complex was completed a week after D-Day. However, the D-Day invasion led to the isolation of the Cotentin Peninsula, and no V-1 equipment was ever deployed at the site. The US Army captured the site in late June 1944. Of the many V-1 heavy sites, it is one of the least known since the launcher is on a French naval reservation and off limits to the public.

bunker was the only one of the original four to nearly reach completion. The reinforced concrete roof entered construction in the last weeks of March 1944 in spite of repeated attacks; Siracourt was struck a total of 27 times with about 5,000 tons of bombs. By the time of D-Day, concrete work was almost complete but only about half of the excavation had been finished since it was necessary to dig out the soil under the roof. The plans called for completion of site mid-July but the launch ramps were never started and indeed, the final configuration of the site is something of a mystery.

A V-1 missile is seen here loaded on the Walter catapult and ready for launch. The gas-generator trolley is in place as is the Anlaßgerät on the left rear side of the fuselage, which contained the electrical controls and air pressure attachments to start the launch. (NARA)

To make up the shortfall of large launch bunkers, in March 1944, the Luftwaffe decided to reinvigorate construction of the proposed Army Ölkeller Cherbourg, a series of tunnels near Brécourt in the Cherbourg suburbs that had been intended for storing A-4 ballistic missiles. The A-4 was not yet ready for production, instead the tunnels would serve as the preparation and storage area for the FZG-76 missiles, while two protected launch ramps were added to turn the facility into a gigantic protected launch site. This new configuration was called Wasserwerk No. 2, and the Brécourt site was variously codenamed the Minenlager (mine storage) or Ersatz B8 since it was intended to replace the ill-fated B8 Waterworks at nearby Couville. The Minenlager was expected to be able to contain 300 FZG-76 missiles, enough for about six days of launches. As expected, the Minenlager attracted far less attention than the more obvious early configurations and even though only 40 percent of the concrete work was ready by mid-May 1944, the work progressed so well that most of the main construction was completed in the week after D-Day. In March 1944, a second FZG-76 site of this type was considered for the Roche de Tronquet tunnels near Nardouet. These tunnels were in use by the Kriegsmarine, but could be modified with four sheltered launchers emanating from the tunnels. It did not progress beyond the design stage.

Operation *Eisbär*

The Allied amphibious invasion in Normandy on June 6, 1944, forced a premature start of the missile campaign against London. At 1745hrs on D-Day, the FR 155(W) headquarters received the codeword Rumpelkammer (junk room). This initiated the transfer of the launch catapults from the main missile depots of Nordpol and Leopold to the operational site system as well as the supply of missiles, fuel and other vital equipment. The plan had been to allot 10 days to this process, but under the circumstances it was shortened to six days. Eight of the new launch sites had to be written off immediately due to their locations in Normandy, and so the four launch battalions operated solely from the Pas-de-Calais area. On the evening of June 12, the orders were given to initiate Operation *Eisbär* (Polar bear), the missile attacks

The V-1 missile was manufactured as easily transportable components that were not assembled until reaching the launch site. The core element consisting of the after fuselage and engine is seen here in the Luftwaffe munitions depot at Dannenberg in April 1945. The fuselage is lacking the warhead, which was only attached at the launch site, and the transport cone on the front protects the nose-mounted air pilot propeller, as well as containing the warhead fuzes. (NARA)

on London. The initial launches were a complete flop. A total of 63 of the 72 launchers were on duty but only nine missiles actually left the launchers in the first salvo, and not one of these reached England. The second salvo, around 0330hrs on the 13th, was little better: ten missiles were launched of which four immediately crashed in the vicinity of the launchers. Two more crashed into the Channel, and four actually reached England, one landing in London in Bethnal Green at 0418hrs. Churchill's science advisor, Lord Cherwell, remarked about the long-dreaded start of the missile campaign: "The mountain hath groaned and given forth a mouse!"

The preparation phase had been too short, and the French road network too disrupted by Allied air attack for the initial preparations to succeed. Many launch battalions were lacking critical parts and supplies, for example IV Abteilung, FR 155(W), lacked the sodium permanganate used to power the steam generator for the catapult. A temporary halt in the operation was

The enormous length of the Walter catapult made them visible to Allied aerial reconnaissance. This example near Amiens has the front section partially collapsed, but it is not clear whether this was due to air attack or by the FR 155(W) crews, who generally sabotaged the launchers before evacuating the launch sites in August 1944. (NARA)

The USAAF favored attacks on production and supply sites rather than the V-1 launch sites. This is a B-17 during a raid on the hydrogen peroxide plant at Peenemünde on September 6, 1944. (NARA)

E Bauvorhaben 21 Felsgestein (Ersatz KNW), Wizernes, France

The chalk quarry outside Wizernes was first selected in 1943 as a supply base for A-4 Feuerteufel (fire-devil) missile operations on the Pas-de-Calais as it could be easily tunneled. This facility was originally codenamed SNW (Schotterwerk Nordwest: Northwest Gravel Works). In the wake of the destruction of the Kraftwerk Nordwest at nearby Watten/Eperlecques, the Wehrmacht needed a new location to create a substitute "Ersatz KNW" and Wizernes was an obvious choice for the facility, which was nicknamed Felsgestein (rock-cliff). As in the case of the Luftwaffe *Wasserwerk* design, the new

Wizernes design, codenamed Bauvorhaben 21, used a new configuration to minimize the risk of air attack. A large 84m (275ft) dome, 5m (16ft) thick was constructed on a hilltop at one end of the lime quarry starting in November 1943. Once complete, the gallery below was excavated and an elaborate launch bunker created. The construction of the main missile launch facility was conducted in parallel to extensive tunneling on the south side of the site along the chalk cliffs. The complex was connected to the main St. Omer–Boulogne railroad line through the Ida tunnel on the north side of the

site. As in the case of the original Watten Bunker, an associated liquid-oxygen plant, codenamed 1302, was also planned consisting of five compressors and a storage capacity of 400 tons.

The main missile preparation gallery under the dome was serviced via the Ida tunnel where trains would deliver missiles, fuel and other supplies. The missiles would then be assembled, erected, and fueled within the protection of the bunker and then dispatched through two tunnels, codenamed Gretchen and Gustav, to a launch plaza on the quarry floor outside. The neighboring tunnel complex would contain additional storage of missiles and missile fuel, and each of the main tunnels were given names such as Katharina, Sophie,

Anna, etc. Besides the main facility, an auxiliary bunker codenamed Umspannwerk C was created nearby in Roquetoire to house a Leitstrahl system. This was a radio command guidance system, which could be used to send course corrections to A-4 missiles launched from Wizernes. Although the bunker was built, it was never put to use. The Wizernes site was bombed several times before D-Day and there was at least one direct hit on the dome during the May 6, 1944, raid which had no effect. The intensity of the Crossbow attacks picked up after the start of Operation *Eisbär*. The RAF began using Tallboy bombs against the site starting on June 24, 1944, and a second Tallboy raid was conducted on July 17, 1944. Although none of the Tallboys managed to impact the dome, a near miss substantially damaged the galleries below, and on July 18, Hitler ordered work at the site to cease. The site endured 16 raids by 811 Allied bombers dropping some 4,260 tons of bombs.

A thoroughly smashed V-1 launch site, presumably after an air attack. The Walter catapult has been broken into several pieces and is being examined by an Allied technical intelligence team. (NARA)

ordered and the attacks resumed two nights later after some of the most egregious supply issues were resolved. During the second attack, on the night of June 15/16, 55 launchers fired 244 missiles, of which 45 crashed after launch, 144 reached the English coast, and 73 fell on London. Seven were shot down by fighters and 25 by anti-aircraft guns. On their own initiative, the 65.Korps also launched 53 missiles against the ports in the Portsmouth–Southampton area, hoping to disrupt Allied naval activity connected with the Normandy operations. When higher headquarters was informed of the port attacks, they were reprimanded for violating Hitler's orders to concentrate on London.

Allied air attacks on the V-1 operational site system June–August 1944							
Date	Bombers	Bombs	Sites destroyed	Serious damage	Medium damage	Slight damage	Casualties*
May 31–June 15	170	650	0	1	0	1	0/0
June 15–30	4,500	18,000	2	22	8	10	20/71
July 1–15	3,300	11,500	2	16	7	14	15/43
July 15–31	1,350	5,700	5	5	6	8	3/14
August 1–15	1,500	6,000	0	11	10	8	18/53
Total	10,820	41,850	9	55	31	41	56/181

*dead/wounded

The start of the "buzz bomb" attacks on London led to a series of hasty attempts to limit or stop the attacks. The attacks caused widespread damage and panic, and Churchill was adamant that the highest priority must be afforded the efforts to stop the missiles, now officially dubbed the "V-1" by Hitler. The most effective methods were entirely defensive: the creation of radar-directed anti-aircraft gun belts along the coast backed by fighters and capped off by a balloon barrage. While this took some weeks to perfect, in the end, some 65 percent of the missiles were shot down. Combined with the high level of technical losses, only about a quarter of the V-1 missiles actually impacted in Britain. The vexing issue of attacking the sites again roiled the leadership of the Allied air forces. On June 16, Eisenhower ordered that attacks on the V-1 sites were to take "first priority over everything except urgent requirements of battle," and they would involve not only the US Eighth and Ninth Air Force, but also the RAF Bomber Command, which had largely avoided participation in the first campaign except for the first Peenemünde raid.

In spite of the diversion of large numbers of heavy bomber attacks against the V-1 sites in the final two weeks of June, there was no appreciable decline in German missile launches. In part, this was due to the problem of actually finding the new sites; of 8,310 sorties in late June, only 4,500 bombed actual

The Allies were baffled by the Mimoyecques site but bombed it thoroughly in the summer of 1944. The plate in the center was the only one to nearly reach completion, but the tunnels below were thoroughly smashed by Tallboy attacks. (NARA)

launch sites. This led to a renewal of the debate about how best to deal with the missile threat. The US Army Air Force was not at all happy to continue to waste large numbers of heavy bomber attacks on the sites, which they regarded as fruitless, and Air Chief Marshall Arthur Harris of RAF Bomber Command largely agreed. The head of the US strategic bomber force in Europe, Lieutenant-General Carl Spaatz, favored attacks against the V-1 production facilities, attacks against more lucrative targets such as V-1 storage dumps, and attacks against the electrical power grid in the Pas-de-Calais that supported the V-1 launch sites. Spaatz also initiated a program to develop improvised guided missiles using worn-out bombers, the Aphrodite program, to attack the heavy sites.

Smashing the heavy sites

The summer Crossbow bombing campaign placed a new emphasis on the heavy sites, in spite of a lack of evidence that they were operational. There was considerable anxiety that they contained even more fearsome new weapons, such as the anticipated V-2 ballistic missile. Previous attacks had mixed results against these bunkers, often able to derail construction plans, but not to eliminate the threat entirely. By July, a new weapon had entered the RAF arsenal that signaled the death knell of the heavy sites—the Tallboy heavy bomb. The Tallboy heavy bombs were the brainchild of Barnes Wallis, the inventor of the bouncing bombs used so successfully against the Ruhr dams in 1943. Two types were built, the Tallboy-Medium weighing 6 tons and the Tallboy-Large weighing 11 tons. The Tallboy (M) was the first to be placed into series production and was first used on June 8, 1944, to close a key railroad tunnel in France. These bombs were so large that they could only be carried by the Lancaster bomber, which had a bomb-bay long enough to accommodate them.

On June 24, Lancasters delivered a dozen Tallboy bombs against the V-2 site at Wizernes that failed to inflict significant damage. Several subsequent attacks were made using more conventional

Operational site system launch rates during Operation *Eisbär* June–August 1944 (daily averages)		
Date	Operating launchers	Missiles fired
June 15–30	39	152
July 1–15	34	119
July 16–30	30	116
August 1–15	34	87
August 16–30	24	72

The nemesis of the heavy missile sites was the new Tallboy bomb, dropped from RAF Lancaster bombers. Appropriately enough, a full-scale model of a Tallboy hangs from the ceiling of the Watten Bunker and a large penetration can be seen in the ceiling. (Author's collection)

bombs, which also failed to damage the main dome. Finally, on July 17, another Lancaster raid with a dozen Tallboys hit the facility, undermining the dome and causing fatal damage. On June 25, 1944, an RAF raid on the Waterworks at Siracourt finally put the site out of commission with a raid that included Tallboy bombs, two of which collapsed portions of the bunker itself. The original heavy site at Watten continued to attract attention by heavy bombers culminating in a July 25 attack by Lancasters and 15 Tallboy bombs, one of which smashed open the rear of the structure. The Tausenfüßler supergun site at Mimoyecques was hit and demolished by Tallboy bombs on July 6, one bomb hitting the gun port slab and three other bombs smashing into the main tunnels. By the end of July, all four of the surviving heavy sites had been effectively crippled by 27 raids consisting of 1,791 bombers dropping some 7,636 tons of bombs. Several of the radio-controlled Aphrodites were used with little success.

Among the most effective Crossbow attacks were those against the two main missile storage facilities, each capable of housing over 1,000 missiles. The Nordpol depot at Nucourt was hit by US Eighth Air Force bombers on June 22 and 24 with 250 tons of bombs followed by RAF Lancasters on July 10–15 with a further 2,165 tons that managed to collapse several of the tunnels and render the depot unusable. The Leopold depot at St. Leu-d'Esserent was hit on the night of July 4/5 with Lancasters carrying 6-ton Tallboy bombs, and several tunnels were collapsed. This was probably the single most effective raid of the summer Crossbow campaign, putting a crimp on German missile supplies for over a week. In the days prior to the attack on the Leopold depot, FR 155(W) was approaching 200 missile launches per day, but after the raid, the totals fell almost by half. These attacks forced the Luftwaffe to disperse the storage areas, and as a result slowed the supply of missiles to the launchers. The FR 155(W) war diary indicates that the main bottleneck on the number of missiles launched were the interruptions in supply of missiles and equipment rather than the attacks on the launchers themselves. Other smaller depots and supply areas were bombed during the Crossbow campaign as they were identified, but the attacks had a much less dramatic effect on dampening down the missile launches.

The other effective attacks of the Crossbow campaign were those by the US Eighth Air Force against the V-1 production sites, notably the Volkswagen plant at Fallersleben with two raids in late June and another in early July. These raids, combined with attacks on the gyroscope plant near Weimar, helped to suppress the scale of V-1 production. The plan had been to raise production to 8,000 missiles by the end of the summer, but the air attacks capped the production at 3,419 in September after which production declined, even after the opening of a second production facility at Nordhausen.

The second phase of the Crossbow raids from June to August 1944 delivered some 95,915 tons of bombs compared to 36,200 tons dropped in the first phase from December 1943 to June 1944. The summer campaign was borne most heavily by RAF Bomber Command, which was often able to operate in daylight due to the destruction of the German fighter force. About 32 percent of the Allied bomb tonnage was directed against the launch sites and, as the senior Allied commanders had warned, these missions were not particularly effective in stopping the missile attacks, destroying only nine launch sites and seriously damaging 55 more. The number of operational launchers did not begin to seriously decline until mid-August when the advancing Allied armies forced FR 155(W) to begin abandoning their launch

areas. The attacks on various types of supply depots and supply sites accounted for about 40 percent of the bomb tonnage. Total V-1 missile production through the beginning of August 1944 was about 9,700 missiles, of which FR 155(W) managed to launch 8,439 (87 percent), which suggests that production problems rather than the Allied attacks on the launch sites were the primary limiting factor on the scale of the summer missile launches.

The Crossbow campaign during the summer of 1944 made it quite clear that mobility and concealment were far better defenses for missile sites than fortification. The heavy sites never became operational due to their vulnerability to air attack. Even without the use of Tallboy bombs, the sites were useless since aerial bombardment could sever the rail and road connections vital to a steady supply of missiles and supplies. In contrast, the

Although the dome of the Wizernes missile base remained largely intact, near misses by Tallboy bombs so undermined the construction that the site was abandoned in the summer of 1944. The small square bunker to the left covered a vent shaft for the missile complex. (Author's collection)

The air-launched V-1 campaign was delayed and an attempt was made to expand the air attacks in the autumn of 1944 to make up for the loss of the French launch sites. The launches were conducted at night, and this illustration shows the launch of a V-1 from an He-111H-22 of KG 3. (Author's collection)

"new" light launch sites proved durable to intense air attack due to the difficulty of locating and identifying them, and the relative ease of repair even when they were bombed. On the other hand, the light V-1 launch sites were not especially efficient as a weapon system as their lack of handling and storage facilities slowed the process of preparing the missiles for launch. The cumbersome Walter catapult took days to disassemble and more than a week to reassemble in another location, making the transfer of batteries from one location to another a difficult and time-consuming process.

Besides the fixed-site V-1 launchers, the Luftwaffe had intended from the outset to use air-launched V-1 missiles as well. The main difficulty of employing aircraft launchers was their poor accuracy due to the lack of sufficiently precise navigation equipment over the North Sea, which amplified the existing accuracy problems of the V-1 missile itself. The mission was assigned to III./KG 3 based in the Netherlands which operated modified He-111H-22 bombers carrying a single V-1 missile under the right wing. The first air-launched V-1 missions began on June 9 and by the time the first wave of attacks ended on September 5, II/KG 3 had launched 300 missiles against London, 90 against Southampton, 20 against Gloucester and 23 at Paris at a cost of two He-111 bombers. About half the air-launched missiles fell within a circle 50 miles around the target, which was about three times poorer accuracy than the ground-launched versions; the RAF thought the missiles launched at Southampton were in fact aimed at Portsmouth. The rapid Allied advance into Belgium and intense Allied air activity over the Netherlands in September 1944 forced the squadron to withdraw into Germany.

Operation *Donnerschlag*

The start of Operation *Cobra* by the US First Army on July 24, 1944, marked the beginning of the Allied breakout from Normandy. Montgomery's 21st Army Group began pushing toward the Pas-de-Calais in August 1944, threatening to overrun the launch sites. On August 9, two of the launch battalions halted operations and began preparations to move, and the following day the remainder of FR 155(W) was warned that missile supply would be interrupted by troop movements as the Wehrmacht began its retreat in northwestern France. There had been some preparation work on new launch sites in western Belgium, so there was some hope the regiment could be reconstituted there. The last launch battalion withdrew from France on August 29, 1944, destroying their launchers before leaving the sites. In the end, only one battalion was able to save their launchers, but about three-quarters of the regiment's troops managed to escape into Belgium. Plans to restart the missile campaign in Belgium faltered due to the rapid Allied advance, which overran the new V-1 launch sites in the first week of September 1944.

Due to a shortage of missiles and launch equipment, only two of the regiment's launch

The V-1 launch sites in Germany were usually located in wooded areas for better camouflage. The US Army captured this site in the spring 1945 campaign. (NARA)

battalions remained under its control while the others were converted back to light flak battalions. It was hoped to convert one of the two battalions to a mobile launch configuration using heavy trucks, but this project was abandoned in October 1944 when it proved impossible to transport the long Walter launch rail system by vehicle. Both launch battalions were deployed back to Germany, with plans to use the missiles against Belgian cities, especially Brussels and Antwerp. New launch sites were scouted along the Rhine in Sauerland and northern Westerwald, but there was some concern about launching near heavily urbanized regions of Germany due to

The Wehrmacht was reluctant to place V-1 launchers near populated areas of Germany due to the large number of failed launches. This V-1 came down prematurely near Rohr in early March 1945, but the warhead failed to detonate, as the safe and arming system had not activated the fuzes yet. (NARA)

the large numbers of missile crashes. Launch sites had already been scouted in the Eifel forests along the Belgian frontier, and the border area had only scattered German villages that were less likely to be hit by wayward missiles. The new launch sites followed the pattern of the new sites in France with minimal construction, and an accent on the use of camouflage to prevent detection and attack. There were some differences in construction of the sites, for example, a more improvised base for the Walter catapult, often using locally available material such as brick for the base instead of the reinforced concrete used in France. The first launches of Operation *Donnerschlag* (thunderclap) began from Germany on October 21, 1944, mostly aimed at Antwerp. An improved autopilot permitted a new "oblique firing" method that allowed the missile to make one turn before aligning itself to the target. The main advantage of this feature was that it made it difficult for the Allies to track the missile launch path back to the launch site. By the end of October, eight launch sites were operational in the Eifel, few of which were discovered by Allied aircraft due to the forest cover in the region and the oblique launch feature. In spite of the best efforts of FR 155(W) to avoid hitting German towns, V-1 missiles often crashed after launch, earning them the grim nickname of *Eifelschreck* (horror of the Eifel).

F FOLLOWING PAGE: V-1 LAUNCH SITE

The V-1 launch site centered around the Walter WR 2.3 Schlitzrohrschleuder catapult. The prerequisites for the site were modest—a concrete platform for attaching the catapult and supporting the gas generator trolley, and concrete pilings for the catapult support. The catapult itself was a modular design in 6m sections usually consisting of eight sections and a muzzle brake at the end for a total length of 49m (160ft). The gas generator trolley was fueled with a volatile mixture of T-stoff (hydrogen peroxide) and Z-stoff (sodium permanganate), which created a pulse of high-pressure gas that pushed a piston down the circular tube at the center of the catapult. This piston was attached to a small frame under the belly of the V-1 missile, accelerating the missile down the ramp where it's pulse jet

engine ignited. The launch was controlled remotely from a nearby bunker via the Anlaßgerät (launch device) mounted aside the rear port fuselage of the missile which included a variety of electrical connectors, safe and arming connections and other necessary triggering devices. It took about 20 minutes to load and launch a V-1 missile, though some well-trained units brought the time down to 18 minutes.

Generally, the launcher was placed near tree lines or within a wooded area to camouflage the launcher from aerial discovery. The launch rail often had camouflage netting draped from either side, though this has been omitted here for clarity. The launcher was delivered in the usual Wehrmacht dark yellow camouflage color, and in some cases additional camouflage was painted on.

F

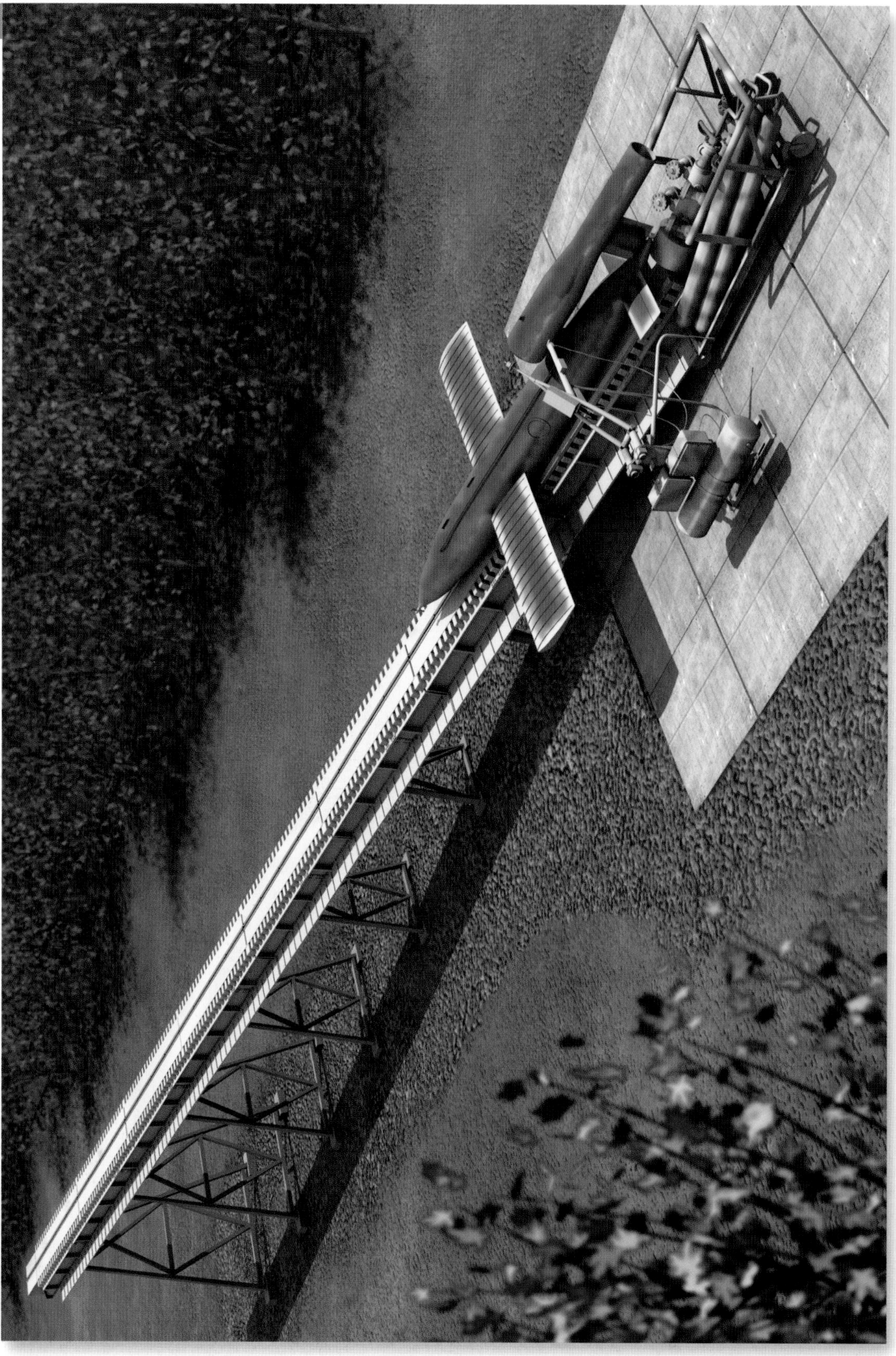

V-1 launch site (caption on previous page)

V-weapons launch sites, Autumn–Winter 1944

Legend:
- V-1 launch area
- V-2 launch site
- V-3/V-4 site
- Main direction of V-1 attack
- Main direction of V-2 attack
- Main direction of V-3 attack
- Major target

0 — 25 miles
0 — 25km

N

London — Rijs
Antwerp

DORA
Dalfsen
Zwolle
Nunspeet
Antwerp
Antwerp
Antwerp
Remagen
Antwerp

The Hague
London
London
ZEPPELIN
Rotterdam
London
XANTHIPPE
Hook of Holland
Antwerp

NETHERLANDS

Antwerp
Antwerp
Antwerp
Munster

London
Walcheren

GERMANY

Dusseldorf
Antwerp

Antwerp
Cologne

Antwerp

Antwerp
Antwerp
Antwerp
Koblenz

Antwerp

Brussels

BELGIUM

Liège

Paris
Eifel

Antwerp

Bitburg

LUXEMBOURG

Antwerp
Hermeskeil
Lampaden
Antwerp
Merzig

FRANCE

Rhine

The V-2 launch site was generally created in wooded areas to help camouflage the launcher, but an extensive road network was needed for the many support vehicles. (NARA)

The difficulties in restarting the V-1 missile campaign from ground launchers led to an expansion of the air-launched effort, with three wings of KG 53 organized for missile launches from bases in the Netherlands and western Germany, with about 100 bombers on strength at the start. The earlier air-launched campaign had gone barely noticed by the RAF due to its small scale, but by mid-September the anti-aircraft gun belt had been extended to deal with air-launched V-1 missiles coming in from the North Sea, and radar-equipped night-fighters were assigned to deal with the missile-bombers. The air-launch missile campaign proved to be ineffective and costly. For example, on a typical night assault on the evening of 16 September, of 15 bombers setting off, only nine released their missiles successfully, three of these were shot down by ships, two more by anti-aircraft and only two reached the London area. Launch failures averaged a quarter to a half of all missiles dropped. By late September, the bombers came under attack by Mosquito night-fighters, losing the first four bombers on the night of September 25, 1944. The campaign dragged on until January 14, 1945, by which time the fuel crisis and heavy losses put an end to any further missions. By the end of the campaign, 1,776 V-1 missiles had been air-launched, of which Allied radars identified 1,012, suggesting that about 700 crashed shortly after launch. Of these, 404 were shot down, including 320 by anti-aircraft fire, 11 by the Royal Navy and 73 by the RAF. Only 388 impacted in England, of which only 66 reached London. A total of 77 He-111 bombers were lost during the attacks, at least 16 to Mosquitoes, and the rest to weather and accidents. In other words, less than four percent of the missiles reached their target and more than one bomber was lost for every missile reaching London, a woefully wasteful option.

The ground-based V-1 campaign was reinvigorated in the late autumn of 1944 as more equipment and troops became available. There were several reorganizations, the original 65.Korps becoming 30.Armee Korps on October 24, 1944. After the SS took control of the V-2 batteries, the corps was disbanded on November 16, 1944, and the Luftwaffe combined FR 155(W) and the partially formed FR 255(W) into 5.Flak Division (W). On November 20, 1944, III./FR 155(W) began Operation *Ludwig*, launching missile attacks on Liège. In the meantime, a third launch battalion was reconstituted with plans to deploy launchers from sites in Germany.

The activation of launch sites in Germany remained contentious. By December, 20 sites had been completed along the Rhine and eight launchers erected, but continued high rates of crashes led to a reluctance to launch from sites near German cites. Instead, the regiment decided to establish new sites in the Netherlands since according to the regimental diary "in Holland there is no need to worry about the civilian population in respect to premature crashes." Two battalions deployed to new sites in the Netherlands around

Deventer, beginning their campaign against Antwerp on 16 December, while III/FR 155(W) continued its attacks from the Eifel against Liège. As in the case of London, anti-aircraft defenses were one of the most effective counter-measures against the V-1, and the US Army began a major effort to defend Antwerp, codenamed Antwerp-X. The launch sites in the Netherlands were a source of great resentment to the Dutch and the resistance fed information to Allied intelligence which led to air attacks. To reduce the vulnerability of the sites, FR 155(W) planned to use the same tactic as in France the previous summer, periodically moving them to avoid detection. This led to Operation *Mülleimer* (dustbin), which moved the launch batteries of II/FR 155(W) to new sites codenamed Xanthippe in the Rotterdam area, and about 300 missiles were fired in eight days starting on January 27, 1945.

The failure of the German Ardennes counter-offensive put the Eifel launch sites at risk and on January 27 III/FR 155(W) joined the other two battalions in the Netherlands at the Dora launch sites for Operation *Oktoberfest*. Since not enough sites were ready there, one battery began launches from the abandoned launch sites around Cologne for a week starting on February 11, 1945. By late February, the supply of fuel and missiles began to decline due to the collapse of the German military industry and the 5.Flak Division was ordered to convert part of its force into an infantry regiment for dispatch to the Eastern Front. In total, FR 155(W) launched 8,696 V-1s against Antwerp, 3,141 against Liège and 151 against Brussels. From October 1944 to March 1945, the V-1 battalions launched 11,988 missiles against Belgian cities of which 1,731 crashed shortly after take-off.

The new extended-range Fi-103E-1 missile became available in February 1945 and it could reach London from launch sites in the Netherlands. A total of 21 launch sites were prepared under the codename Zeppelin for Operation *Pappdeckel* (pasteboard). The attacks began on March 3, 1945, and 275 were launched against London through to March 29, 1945. Of these, only about 160 flew any significant distance, 92 were downed by air defenses and only 13 reached London, the last on March 28, 1945. With their launch sites about to be overrun in the Netherlands, the missile campaign came to an end. In total, the V-1 attacks had killed about 5,500 and wounded 16,000 in England as well as causing substantial damage.

The V-2 missiles were usually shipped to the launch areas by railroad, transferred to a Vidalwagen transporter and then finally loaded on to the Meillerwagen transporter-erector using a 15-ton Strabo overhead crane. (NARA)

G

V-2 launch site

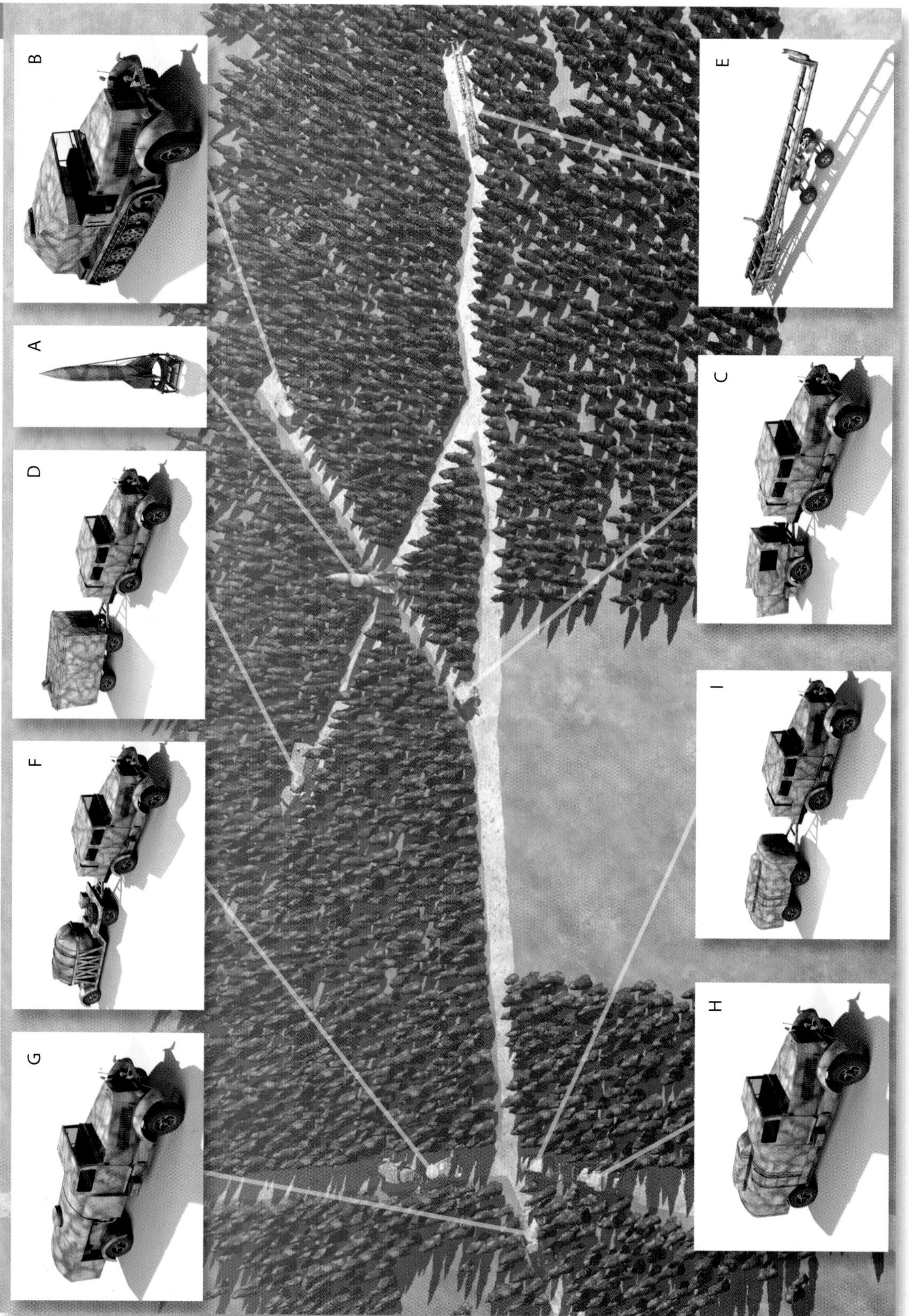

The Meiller launch pad was towed to the launch site, in this case behind the SdKfz 7 8-ton half-track armored fire-control vehicle. The wheeled trailer was removed before the missile was placed on the pad. (NARA)

Operation *Pinguin*

The new A-4 missile debuted in the autumn of 1944 after long delays, renamed by Hitler as the V-2. Three missile launch battalions had been formed in late 1943, Artillerie Abteilungen 485, 836 and 962 (Mot.). In the spring of 1944, SS-Werfer Battalion 500 began converting from their conventional artillery rocket launchers to the A-4 as part of Himmler's efforts to place the SS in control of the new secret weapons. The original scheme was to deploy these in Normandy and the Pas-de-Calais at heavy sites such as Wizernes and Sottevast, as well as from mobile launch sites. A-4 missile production began at the damaged Peenemünde plant in late 1943, but a new production facility, codenamed Mittelwerke, was created in the Harz Mountains near Nordhausen by tunneling under a mountain. Production began in the tunnels there in January 1944 using slave labor from the notorious Dora camp nearby. However, the A-4 was plagued with technical

G V-2 LAUNCH SITE

The V-2 launch site was located in an area about 500m (1,600ft) in length and this shows the textbook configuration. The prescribed deployment pattern was in a wooded area, or a road network edged with trees to help camouflage the launch site from aerial observation. The launch area itself was ideally a flat clearing about 50m (165ft) wide with good access to roads. The firing platform (A) was located at the center of this clearing with three key pieces of equipment nearby: the battery fire-control vehicle (B), electrical generator trailer (C) and an air compressor trailer (D). Each was positioned about 90m (300ft) from the launch pad, as there was the constant danger that the missile engine turbopump would fail shortly after take-off, with the missile and its fuel crashing down on the pad and exploding. The fire-control vehicle, based on a Krauss-Maffei SdKfz 7 half-track, was deployed with a clear line of sight to the launch pad since the battery commander conducted the launch from this site. The generator and compressor trailers provided electrical power and hydraulic air pressure to the missile and launcher

and the associated support equipment. Usually the Meillerwagen tranporter-erector trailer and Magirus servicing ladder were parked near the launch pad as well (E). The fuel equipment was kept further away from the launch pad, usually 365m (1,200ft) due to the dangers of the launch. The A-Stoff (liquid oxygen) detachment consisted of an Anhanger 6 insulated liquid oxygen trailer and its tractor, usually a Hanomag SS-100 (F). The T-stoff (hydrogen peroxide) detachment usually consisted of a Opel Blitz Kessel-KW.2100 1 tanker truck towing a support trailer used to heat the hydrogen peroxide (G). The B-stoff (alcohol) fuel detachment was the largest of these units, usually consisting of an Opel 3-ton KW Kfz. 385 Kessel-KW.3500 1. tanker (H); a Kessel-KW.3500 1. trailer along with its Hanomag SS-100 tractor (I), and a fuel pump trailer. These fuel vehicles would be driven to the launch pad only after a missile had been erected, and would depart as soon as the fueling process was complete for security reasons.

After the missile was erected on the launch pad, the V-2 service battery deployed a variety of fuel trucks, pump trailers and other support equipment to fuel the missile, as seen in this test launch conducted for the Allies in 1946. (NARA)

problems, disintegrating in the final stages of flight, and these problems were not resolved until the summer of 1944. In spite of the obvious failure of the heavy sites, there were plans to create three more fortified bunkers for the V-2 in the western area of Germany, but these plans never came to fruition due to the resistance of the Army, which saw the V-1 experience in the summer of 1944 as confirmation of the tactical benefits of mobile versus fortified launchers. As a result, all V-2 operations in 1944–45 were based on the mobile launcher configuration.

Each V-2 launch battalion consisted of five sections: a headquarters section, launch section, radio section, technical section and fuel section. The launch section had three launch batteries each with three Meillerwagen transporter-erector trailers, one Bodenplatte launch pad and one armored fire-control vehicle. Each launch battery had 39 troops in five teams: fire control, survey and adjustment, engine, electrical, and vehicle/trailer team. The radio section was responsible for unit communications, and conducting the site survey to locate the launch batteries. The technical section was responsible for unloading missiles from the railway supply point, and preparing and transporting them to the launch site. The fueling section was divided into three teams, handling liquid oxygen (LOX), alcohol, and sodium permanganate (Z-stoff) for the rocket motor turbopump. The battalion had an extensive array of fueling vehicles including 22 LOX trailers, 48 alcohol tank trucks/trailers, four hydrogen peroxide trailers and four pump trailers.

The process of preparing a V-2 missile for launch took four to six hours, of which the final 90 minutes was spent actually erecting and fueling the missile at the launch site. The missile was loaded on a Meillerwagen without fuel and towed to a pre-surveyed launch site, ideally a road with trees on either side to provide camouflage during the lengthy launch process. On arriving at the launch site, it took 12 minutes to erect the missile on to its launch pad, at which point the rest of the battery's 32 vehicles and trailers moved into position around the site to prepare and fuel the missile. Fueling took about ten minutes, during which time checks were performed on various missile subsystems. Final checks were conducted after the fueling was completed, and the battery troops and vehicles withdrew a safe distance from the launch site. The launch command was given by the battery commander in an armored Feuerleitpanzer fire-control vehicle, an SdKfz 7 half-track with an armored shelter on the rear, 100 to 150m from the missile launch pad, behind a protective berm if time permitted. Once the missile was launched, the battery would usually relocate some distance away in case Allied aircraft had observed the launch. About 15 percent of the missiles failed to leave the pad, most often due to technical problems caused by the super-cold liquid oxygen, and even those missiles which did get off the pad often experienced failures in the first few minutes of flight. The launch failures were a significant threat in the Dutch towns from which the missiles were launched since they often fell into neighboring towns and exploded with a full load of fuel. The random destruction in The Hague became so bad that German civil authorities recommended halting the launchings from the city, which was ignored by the SS commander of the V-2 batteries, Brigadeführer Hans Kammler.

The first unit in combat was training Batterie 444, which deployed in the Belgian Ardennes in early September 1944 to carry out Hitler's orders to destroy Paris after if had been abandoned without a fight in late August. The first V-2 combat launch occurred on September 8 against Paris, but it apparently disintegrated. A second missile launched later in the morning struck southeast of Paris, killing six people and injuring 36. This ended the V-2 attacks on Paris, as Hitler was adamant that the weapon be focused against London. A battery of Artillerie Abteilung 485 began operations from The Hague on September 3 against London and two missiles impacted there in the early evening of the same day. The attacks continued at a slow rate and another battery arrived in The Hague on September 10. The launch rate was limited by the supply of liquid oxygen, and by the poor technical state of the missiles. Two batteries of Artillerie Abteilung 836 went into action from Euskirchen on September 15, mainly aimed at cities in France such as Lille. On September 16, Batterie 444 joined the bombardment of London after moving to the coast near Walcheren. During the first phase of Operation *Pinguin* (Penguin), a total of 43 V-2 missiles were launched: 26 against London and 17 against other cities, mainly in France.

Unlike the V-1, there was little the Allies could do to counter the V-2 missile. The missile launchers moved after firing, and neither anti-aircraft guns nor fighter aircraft were effective against a ballistic missile. The V-2 batteries were temporarily disrupted by Operation *Market-Garden*, the Allied airborne operation in the Netherlands in mid-September, but sporadic firings resumed later in the month when two batteries returned to The Hague to continue the attacks on London. During the second phase of Operation *Pinguin*, a total of 162 A-4 missiles was launched of which 52 were aimed at England. The average launch rate increased to 6.5 missiles per day. On October 12, Hitler ordered that the units stop wasting their missiles on secondary targets. Instead, the batteries were to concentrate their attacks on London and the vital port city of Antwerp.

During October, the SS took over control of the V-2 missile program, placing the batteries under the command of Kammler's Division zbV (Division zur besonderen Verwendung: special purpose division). Its Gruppe Nord consisted of Artillerie Abteilung 285 and SS-Werfer Batterie 500, divided between the Burgsteinfurt area in Germany targeting

The Meillerwagen erector frame served as a gantry for servicing the missile after it had been erected on the launch pad behind it. It was towed away immediately prior to launch. (MHI)

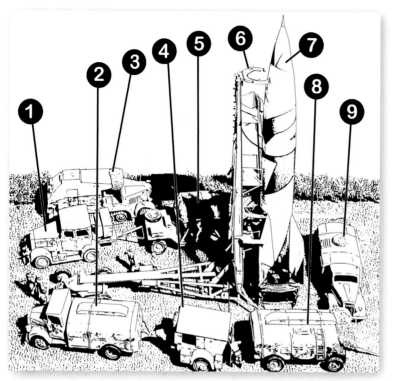

This illustration shows a typical launch site during the critical fueling process including the
(1) Hanomag SS-100 tractor;
(2) Opel 3-ton KW Kfz. 385 B-stoff (alcohol) Kessel-KW.3500 1. tanker;
(3) SdKfz 7 8-ton half-track armored fire-control vehicle;
(4) fuel pump trailer;
(5) A-stoff (liquid oxygen) Anhanger 6 insulated trailer;
(6) Meillerwagen transporter-erector;
(7) A-4 ballistic missile;
(8) B-stoff (alcohol) Kessel-KW.3500 1. trailer;
(9) Opel Blitz Kessel-KW.2100 1. T-stoff (hydrogen peroxide) tanker. (Author's collection)

Antwerp, and The Hague targeting London. Gruppe Sud was smaller, based around Artillerie Abteilung 836, located mainly in the Merzig area to target Antwerp. December 1944 saw an escalation of the launches against Antwerp and neighboring Belgian cities such as Liège coinciding with the German offensive in the Ardennes since Antwerp was the ultimate objective of the German operation. From December 14 to January 4, an average of 100 missiles per week fell on Antwerp. By the end of 1944, a total of 1,561 A-4 missiles had been launched, of which 491 (31 percent) had been aimed at Britain, 924 at Antwerp (51 percent), and the rest at various cities in France and Belgium.

The month of February 1945 was one of most intense phases of A-4 launches, many conducted from the Duindigt racetrack in The Hague. The use of urban launch sites created a significant dilemma for the Allies, who were unwilling to use the sort of carpet-bombing attacks that had been conducted against the V-1 sites in rural France in the summer of 1944. There was an understanding between the British and Dutch governments to minimize civilian casualties even after the RAF learned from Dutch resistance sources that LOX was being supplied to the missile batteries by eight Dutch plants, located near residential areas. Instead of bomber attacks, more than 10,000 fighter sorties were flown against rail and road networks near The Hague and Hook of Holland areas to disrupt missile supplies. The use of the Duindigt racetrack area finally became so intolerable that it was heavily bombed in early March 1945, finally forcing the German missile batteries to abandon the area. The risks of bombing in urban areas was made painfully clear on the night of March 3/4 when the RAF attempted to strike a V-2 storage area in the Hague City Forest but instead dropped 86 tons of bombs on the Bezendenhout suburb, killing more than 500 Dutch civilians and making 30,000 homeless.

Of the 1,359 A-4 missiles launched against London, 1,039 were launched from The Hague and its suburbs while the rest were launched mainly from the Hook of Holland area. Of the 1,359 A-4 missiles launched against London, 169 (12 percent) were failures shortly after launch, 136 (10 percent) disintegrated in the terminal phase of the flight, 1,054 actually reached England, and only 517 hit the city and its suburbs, about 38 percent of the V-2 missiles launched. Civilian casualties in Britain caused by A-4 attacks totaled 2,754 dead and 6,523 wounded. Although London was the best known target of the V-2 attacks, Antwerp in fact sustained more attacks, totaling some 1,610 launches, of which only 598 fell into the city itself (37 percent) and only 152 into the main target, the city's harbor. The launch area for most of the Antwerp attacks was in the Eifel area of Germany and Belgium, although some missiles were launched against Antwerp from the Dutch sites. The British Rhine offensive in March 1945 finally put an end to A-4 launches from the Netherlands. The last six missiles were launched from The Hague against London on 27 March and the last against Antwerp the same day from Hellendoorn. There were plans to move Artillerie Regiment 901 north of Hanover for the "Blücher Mission," a missile strike against Red Army forces around the encircled fortress of Kustrin. The US Army captured the Nordhausen production plant on April 10, 1945, though V-2 production had petered out in March 1945.

The V-2 mobile launch system had proven to be resistant to air attack due to its mobility, but not especially efficient due to the complexity of the missile and its associated systems.

The V-3 and V-4 in action

Although the Tausenfüßler supergun was not ready in time to be fitted to the Wiese Bunker at Mimoyecques, development work continued on the concept through the autumn. The prospects for such a long and cumbersome weapon were poor, so a shortened design was developed called the LRK 15 F 58 (LRK: Langrohrkanone: long-barreled gun), also nicknamed the Fleißiges Lieschen (Busy Lizzie). Instead of fitting this weapon inside a bunker, it was designed to be surface mounted on a suitable hill, much like the HDP test guns at Misdroy. As in the case of the V-2 missile, Artilerie Abteilung 705 was taken over by Kammler's Division zV in the autumn of 1944 when the SS usurped control over the vengeance weapons. Since the gun in its shortened form was incapable of reaching London, it was planned to use it against other targets, in support of the Ardennes offensive in December 1944. A suitable hill was selected in Lampaden near Trier, with the target being Luxembourg City, about 45km away. The Organization Todt was dispatched to clear the site of trees and to prepare the slope to accommodate two HDP guns. Construction of the first gun was completed on December 28 and the second two days later, in time for the arrival of the first supplies of

The Rheinbote rocket was launched from a simple rail mounted on a V-2 Meillerwagen transporter-erector fitted with a blast deflector at the rear. As a result it had no precise traverse, degrading the already poor accuracy of the weapon. (NARA)

ammunition. Five rounds were fired into Luxembourg on December 30, and in total some 183 rounds were fired through February 22, 1945. The main impediment to the more extensive use of the HDP guns was the shortage of ammunition. The guns did not prove to be especially effective, and of the 142 rounds that impacted in Luxembourg, casualties amounted to 10 dead and 35 wounded. The US Army was aware that some sort of long-range gun was being used, but aerial reconnaissance could not find them as they were well camouflaged on the Lampaden Hill. The battery was withdrawn in late February due to the advance of the US Army into Germany.

A second battery of HDP guns began deployment in January 1945 at Bühl in the Vosges Mountains, aimed at Belfort to support the Operation *Nordwind* offensive in Alsace. Although one gun was erected at the site, the failure of the *Nordwind* offensive put the site at jeopardy and the equipment was withdrawn before firing began. There were other schemes to deploy batteries to bombard Antwerp and other cities, but these came to naught owing to the disruption of the German railroad networks by Allied air attack and the lack of the specialized ammunition. All four HDP guns were dumped at the Röchling plant in Wetzlar, and Artillerie Abteilung 705 was reorganized with conventional guns.

The last of the V-weapons to enter service was the Rheinbote artillery rocket, sometimes dubbed the V-4. This rocket was developed by Rheinmetall-Borsig starting in 1941, and used a multistage solid rocket. It had no guidance beyond fin stabilization. The program received very little priority, even though it had an impressive 150km range, because the warhead was only 40kg with a 25kg high-explosive fill. The project was saved from obscurity in 1944 due to the growing interest of Himmler and the SS in long-range vengeance weapons. In November 1944, the Rheinbote was demonstrated to Brigadeführer Kammler, who ordered the production of 300 of the Rh-Z-61/9 rockets for his Division zV. The test unit was reorganized as Artillerie Abteilung 709 under the command of Oberstleutnant Alfred Tröller and dispatched to Nunspeet in the Netherlands in late December 1944 with plans to join in the bombardment of Antwerp.

The launcher for the Rheinbote was an FR-Wagen, a modified version of the Miellerwagen transporter-erector used for the V-2 ballistic missile. A launch rail was fitted on the erector frame, and the launcher elevated to a suitable angle for launch rather than being placed on a separate launch pad as was the case with the V-2 missile. Artillerie Abteilung 709 was supposed to be equipped with 12 FR-Wagen but only four launchers were available when the first launches were conducted on Christmas Eve 1944. About 45 rockets were launched against Antwerp through the middle of January 1945, but there is no record of their impact and Allied intelligence was largely unaware of the weapon. By early February, even Kammler realized that the rocket was useless, and the program was cancelled on February 6, 1945.

THE MISSILE SITES IN RETROSPECT

The German missile campaign against British and Belgian cities in 1944–45 was the first large-scale use of guided missiles in history with some 23,172 V-1 and 3,172 V-2 missiles launched. The missiles failed to have any decisive effect on the outcome of the war and in April 1945 even Hitler admitted that they had proven to be a total flop. The total tonnage of missiles impacting in London in nine months of attacks was comparable to a single Allied bombing raid of the time. In comparison, the diversion of resources was tremendous, estimated as costing about $3 billion, or about triple the cost of the US atomic bomb program. Another assessment concluded that the resources were comparable to the production of 24,000 fighter aircraft. Of the two missiles, the V-1 was clearly the more effective as it proved simpler to use under actual combat conditions. For example, about 71 percent of the V-1 missiles manufactured were actually launched while only about 49 percent of the V-2s were launched, in no small measure due to the difficulty of supplying liquid oxygen to the launch sites. Furthermore, the US Strategic Bombing Survey concluded that the V-1 campaign was disproportionately costly to the Allies due to the extensive costs of the countermeasures such as the diversion of bombing missions and anti-aircraft forces to combat the threat. This may be the case, but the Allies could easily afford such costs, while it is doubtful that the missile program was a wise investment for the overstretched German military economy.

The missile campaign had some implications for future missile use. The use of mobile missile launchers was clearly more effective during the course of a long campaign than fixed sites, no matter how fortified. On the other hand,

The HDP gun was set up on a hill near Lampaden for firing against Luxembourg City. This shows the original test configuration at Misdroy near the Baltic, but the combat site was similar. (MHI)

the German mobile missile launchers had been developed in haste and were extremely inefficient even if more survivable. For example, the original V-1 site was expected to have a maximum rate of fire of 72 missiles per day but the mobile site had a rate of fire of less than four per day when in actual service. When the US began producing a copy of the V-1 as the Loon missile, they immediately replaced the cumbersome rail catapult with a much more versatile rocket-assisted take-off (RATO) system that made the system truly mobile. The V-2 missile suffered from its use of cryogenic fuel that limited its field deployment due to the technical and logistical complications of liquid oxygen. Although several armies after the war fielded tactical ballistic missiles with cryogenic fuel, all found them to be too inefficient under field conditions until more practical hypergolic and solid fuel alternatives became available. The classic example of this was the Soviet Army, which manufactured copies of the V-2 as the R-1 and R-2 missiles, but which did not begin to field significant numbers of tactical ballistic missiles until the advent of the R-11 (Scud) missile in the late 1950s, which used the newer and more convenient hypergolic fuels.

Even if fortified sites proved impractical under prolonged combat conditions, there was a major revival of the concept in the Cold War with strategic missile systems. All early intercontinental ballistic missiles used fixed sites, if for no other reason than that the missiles were too complicated and too large to operate from mobile launchers. With the advent of nuclear weapons, these sites went underground using hardening techniques that made them far more survivable than the German missile bunkers of World War II so long as opposing intercontinental missiles were not especially accurate. However, as ICBMs became more accurate by the 1970s, fixed silo sites began to fall out of favor due to their vulnerability, and there was a return to mobile basing, including submarines, truck-mobile launchers and even rail-mobile launchers.

The most curious echo of the 1944–45 missile campaign occurred nearly a half-century later in the Middle East. The Iraqi armed forces turned to ballistic missiles and other exotic weapons due to their inability to field a competent strike aviation force, a similar reason to the Luftwaffe case in 1944. The Iraqi missile force deployed both fixed Scud launchers and mobile Scud launchers during the Operation *Desert Storm* air attacks in 1991, the fixed sites were destroyed almost immediately while the mobile Scud launchers proved a nettlesome if indecisive threat for the whole campaign. And strangely enough, the Iraqis also built a supergun patterned after the German Tausenfüßler, which proved to be every bit as useless. The one major difference between the German and Iraqi cases was the lack of an Iraqi cruise missile arsenal, arguably the more effective of the German vengeance weapons.

THE SITES TODAY

Most of the major Crossbow heavy sites remain today, and three of them have become museums that are worth a visit. The Wizernes Bunker, popularly known in France as "La Coupole" because of its domed bunker, has been converted into an excellent museum with many interesting displays including a V-1 and V-2 missile. The neighboring Watten Bunker does not have as many displays, but the ruined half of the building has been left in its bombed condition and several of the large bomb craters are still in place, giving some idea of the intensity of the air campaign against these sites. British engineers demolished the Mimoyecques Bunker for the V-3 supergun in 1945, destroying the upper gun plate. After the war, the tunnels were gradually cleaned out and the site used for a time for mushroom production. In recent years, the site has been converted to a museum and is very impressive if only for the sheer length of its tunnels. All three of these sites are in the Pas-de-Calais area and can be visited in a single day. Other sites are still in existence but unrestored or inaccessible. The Brécourt Wasserwerk No. 2 launcher

A number of V-2 missiles are preserved in museums around the world, but only a few include the launch equipment. This is the restored V-2 at the US Air Force Museum at Wright-Patterson AFB in Ohio painted in wartime camouflage colors. (USAF Museum)

remains, but is on a French naval base and requires permission from the Cherbourg Naval Command for a visit. The *Wasserwerk* at Siracourt is on private land though still visible from the road, and other sites such as Sottevast are not readily accessible.

The numerous V-1 sites through France and Germany have left behind a remarkable number of small structures, especially the initial "old-pattern" sites. These sites are most often on private land, and in many cases the buildings are so nondescript that a detailed guide is essential. Several of the books listed overleaf provide a guide to these sites in several regions. There are a handful of preserved sites, the best known of which is the former FSt No. 685 at Le Val-Ygot near Dieppe which is now enclosed in a wooded area at the edge of the Forêt d'Eu. The site was heavily bombed but many of the major structures associated with an "old-pattern" site remain. There are other preserved sites at Bois-des-Huit Rues, Yvrench-Bois-Carré and Bachimont, the latter of which is one of the few preserved "modified" sites.

Several sites in Germany with a connection to the V-weapons have museums. The Dora slave labor camp at Nordhausen has been a memorial since 1964, but since German re-unification the site has expanded and some of the tunnels of the Mittelwerke cleared out for visiting; the accent in these exhibits focuses more on the hellish conditions of the Dora camp than on the missile program. The Peenemünde test site has had a more elaborate museum created in recent years, and the site still has some of the structures from the proving ground though the V-2 missile on display is a replica. There are some remains of the V-1 launch sites in western Germany, but few of these have been systematically preserved and they are very difficult to find.

V-weapon artifacts have been widely preserved, and many V-1 and V-2 missiles remain in major aerospace museums in Germany, France, the US and Britain. The Imperial War Museum at Duxford has an exceptional exhibit on the V-1 that contains the most complete set of launch equipment including

V-1 missiles can be widely found at aviation museums around the world, but the launch equipment is more difficult to find. The most thorough collection is at the Imperial War Museum, Duxford, but other museum, such as the Watten Bunker at Eperlecques, France, have partial Walter catapults like this one. (Author's collection)

a Walter catapult ramp, as well as the associated steam generator and electrical launch apparatus. A few restored V-2 missiles on their Meillerwagen transporter-erector remain, including one at the US Air Force Museum at Wright-Patterson AFB, Ohio, and the Australian War Memorial in Canberra. The Battle of the Bulge Museum in Diekirch, Luxembourg, has a preserved example of a Tausenfüßler projectile and there is another at the US Army Ordnance Museum at Aberdeen Proving Ground, Maryland. RAF Museum Cosford has a rare example of the V-4 Rheinbote rocket.

FURTHER READING

The German V-weapons have been the subject of numerous books and studies and this bibliography is by no means exhaustive. The Hautefeuille book is by far the most thorough study of the heavy V-weapons sites while the Delefosse book is the most thorough study of the V-1 and includes very detailed drawings and photographs of the many types of buildings associated with the launch sites. Three of the books provide an especially detailed regional survey of V-1 launch sites: Bailleul (Picardie, Artois, Flanders); Grenneville (Normandy); and Gückelhorn/Paul (Germany). The articles listed here are those with a special focus on the launch sites. Those in *After the Battle* magazine are especially helpful for anyone planning to visit the sites. The Operation *Backfire* report is a multi-volume study completed after British forces conducted experimental V-2 launches with the assistance of former German missile troops in 1946 and provides an extremely detailed look at the process of preparing and firing a V-2 missile as well as a considerable amount of detail on the equipment and organization of German ballistic missile units.

There are numerous wartime intelligence reports on the V-1 and the author referred to collections at the US National Archives, US Army Military History Institute, and the Smithsonian's National Air and Space Museum. An exceptional resource on the V-2 is the V-2 Rocket Internet site (www.v2rocket.com).

Government studies
Handbook on Guided Missiles: Germany and Japan (US War Department, 1946)
Investigation of the Heavy Crossbow Installations in Northern France (3 volumes, Sanders Mission, Crossbow Committee, 1945)
Helfers, Lt. Col. M., *The Employment of V-Weapons by the Germans During World War II* (US Department of the Army)
Kriegstagebuch Flak Regiment 155 (W) (English Translation in Imperial War Museum, London)
Report on Operation Backfire, (three volumes, The War Office, London, 1946)
Walter, Gen. Eugen, *V-Weapon Tactics (LXV Corps)* (US Army Foreign Military Studies B-689, 1947)
V-Weapons (Crossbow) Campaign (US Strategic Bombing Survey, 1945)

Articles
Ehmke, Axel, "V1 Endfertigungs-Lager und Abschußanlage in Brécourt," *DAWA-Nachrichten*, No. 30, 1997, pp. 29–41
Ehmke, Axel, "Wasserwerk Nardouet- Planungen für ein V1 Abschußanlage," *DAWA-Nachrichten*, No. 32, 1998, pp. 33–36
Garbe, Horst, "V-Waffen im Rheinland," *DAWA-Nachrichten*, No. 23/24, 1994, pp. 11–16

Held, Michael, and Lippmann, Harry, "V1- Abschußrampe nordlich Kuchem/Sieg," *DAWA-Nachrichten*, No. 27, 1996, pp. 9–12

Heitmann, Jan, "The Peenemünde Rocket Centre," *After the Battle*, No. 74, 1991, pp. 1–25

Magry, Karel, "Nordhausen," *After the Battle*, No. 101, 1998, pp. 2–43

Pallud, Jean-Paul, "HDP: Le bunker de Mimoyecques," *'39/45 Magazine*, January 2003, pp. 44–58; May 2003, pp. 44–53

Pallud, Jean-Paul, "La Rheinbote ou V-4," *'39-45 Magazine*, April 2004, pp. 40–52

Pallud, Jean-Paul, "The Secret Weapons: V3 and V4," *After the Battle*, No. 114, 2001, pp. 3–27

Speth, Ronald, "Visiting the Mittelwerk: Past and Present," *Spaceflight*, Vol. 42, March 2000, pp. 113–19

Thiele, Olive, "Unternehmen Rumpelkammer: Eine Chronik der V-1 Abwurfe aus der Luft," *Flugzeuge*, No. 1, 1988, pp. 36–42

"The V-Weapons," *After the Battle*, No. 6, 1974, pp. 2–41

Books

Bailleul, Laurence, *Les Sites V1 en Flandres et en Artois* (Self-published, 2000)

Bailleul, Laurence, *Les Sites V1 en Picardie* (Self-published, 2006)

Cuich, Myrone, *Armes Secrete et Ouvrages Mysterieux de Dunkerque a Cherbourg* (Tourcoing, 1984)

Darlow, Steve, *Sledgehammers for Tintacks: Bomber Command Combats the V-1 Menace 1943–44* (Grub Street, 2002)

Delefosse, Yannick, *V1-Arme du désespoir* (Lela Presse, 2006)

Dornberger, Walter, *V-2* (Viking, 1954)

Ducellier, Jean-Pierre, *La guerre aerienne dans le nord de la France: 24 Juin 1944, V-1 Arme de Représailles no. 1* (Doullens, 2003)

Dufour, Norbert, and Doré, Christian, *L'enfer des V-1 en Seine-Maritime durant la Seconde Guerre Mondiale* (Ed. Bertout, 1993)

Dungan, T. D., *V-2: A Combat History of the First Ballistic Missile* (Westholme, 2005)

Glass, A. , et. al, *Wywiad Armii Krajowej w walce z V-1 i V-2* (Mirage, 2000)

Grailet, Lambert, *Liège sous les V-1 et V-2* (self-published, 1996)

Grailet, Lambert, *Le V-3 harcèle Luxembourg* (self-published, 1996)

Grenneville, Regis, *Les Armes Secrètes Allemandes: Les V1* (Heimdal, 1984)

Gruen, Adam, *Preemptive Defense: Allied Air Power Versus Hitler's V-Weapons 1943–45* (USAF, 1998)

Gückelhorn, Wolgang, and Detlev, Paul, *V1-Eifelschreck* (Helios, 2004)

Hautefeuille, Roland, *Constructions Speciales* (Tourcoing, 1995)

Hellmold, Wilhelm, *Die V1: Eine Dokumentation* (Bechtermünz Verlag, 1999)

Henshall, P., *Hitler's V-Weapons Sites* (Sutton, 2002)

Hölsken, D., *V-Missiles of the Third Reich* (Monogram, 1994)

Irving, D., *The Mare's Nest* (Wm. Kimber, 1964)

Kennedy, G., *Vengeance Weapon 2* (Smithsonian, 1983; Shiffer reprint, 2006)

Klee, E., and Merk O., *The Birth of the Missile* (Harrap, 1965)

Laskowski, Piotr, *Niemieckie tajne bronie na wyspach Wolin i Uznam* (Maagdruk, 1999)

McGovern, J., *Crossbow and Overcast* (Wm. Morrow, 1964)

Thompson, Peter, *V3: The Pump Gun* (ISO Publications, 1999)

Verbeek, J. R., *V2-Vergeltung: From The Hague and its Environs* (V2 Platform Foundation, 2005)

Young, Richard A., *The Flying Bomb* (Ian Allen, 1978)

INDEX

Figures in **bold** refer to illustrations.